T0413933

EARTHQUAKES AND TSUNAMIS

GEOTECHNICAL, GEOLOGICAL AND EARTHQUAKE ENGINEERING

Volume 11

For other titles published in this series, go to:
http://www.springer.com/series/6011

Earthquakes and Tsunamis

Civil Engineering Disaster Mitigation Activities

Implementing Millennium Development Goals

edited by

A. TUGRUL TANKUT

Middle East Technical University, Ankara, Turkey

Editor
A. Tugrul Tankut
Department of Civil Engineering
Middle East Technical University
0651 Ankara
Turkey
ttankut@metu.edu.tr

ISSN 1573-6059
ISBN 978-90-481-2398-8 e-ISBN 978-90-481-2399-5
DOI 10.1007/978-90-481-2399-5
Springer Dordrecht Heidelberg London New York

Library of Congress Control Number: 2009926872

Printed on acid-free paper

Springer is part of Springer Science+Business Media (www.springer.com)

Preface

The present volume consists of eleven keynote papers presented at the Earthquake and Tsunami conference jointly organised by the World Council of Civil Engineers (WCCE), the European Council of Civil Engineers (ECCE) and the Turkish Chamber of Civil Engineers (TCCE) to be held on 22–24 June 2009 in Istanbul, Turkey.

Natural disasters have always been an area of prime concern for all of the three organising institutions. The proposal of TCCE to jointly organise a conference on earthquake and tsunami was therefore enthusiastically accepted by WCCE and ECCE. The event was realised by the Organising Committee comprising professors from the leading Turkish universities and TCCE representatives. The International Advisory Board consisting of the representatives of the three organising partners discussed and set the organisational principles. It was unanimously agreed by both the Board and the Committee to set the scientific standard of the conference through well qualified, internationally renowned keynote speakers. Each of the six half-day sessions would be initiated by two keynote presentations relevant to the subject area of the session.

The objective of the conference is to contribute to the mitigation of life and material losses in earthquake and tsunami through improved civil engineering practice. The conference scope is confined to civil engineering related disaster mitigation activities. Consequently, the keynote papers included in the present volume, deal with disaster mitigation in various areas of civil engineering.

Although detailed information is available in their respective web pages, it would be appropriate to briefly introduce the three institutions organising the conference.

Turkish Chamber of Civil Engineers (TCCE) – is a big NGO that has more than 70 000 registered members. Its primary objectives include betterment of the civil engineering practice in the country and improvement of the employment conditions of its members. To this end, it runs the professional engineering system; it provides continuing education courses, exams and credits; it issues private practice licences; it organises professional and scientific symposia in collaboration with the academia; it represents the Turkish civil engineering community in international circles. It also takes interest in the social and economic developments in the country and responds to them by expressing opinion and proposing alternatives.

European Council of Civil Engineers (ECCE) – brings civil engineering organisations from the European countries together. One organisation represents each member country in ECCE. ECCE aims to promote the highest technical and ethical standards and to provide a source of impartial advice. To this end, ECCE advises individual governments and professional institutions, formulates technical and ethical standards and guidelines to maintain and raise standards of civil engineering education, training and professional competence, as well as encouraging and improving levels of safety and quality. ECCE activities are currently carried out by five standing committees; namely, Professional Recognition & Mobility; Education & Training; Knowledge & Technology; Environment & Sustainability; Business Environment.

World Council of Civil Engineers (WCCE) – provides a global platform for various entities of civil engineering. National and international organisations, construction companies and even individuals can be accepted as members. Its objectives include promoting high technical and ethical standards of civil engineering, combating corruption worldwide, responding to the challenges of sustainable development, solidarity with developing countries, fostering cooperation with other international organisations of civil engineering. WCCE activities are currently carried out by four standing committees; namely, Natural Disasters; Construction; Education & Training; Water Resources.

The present volume is arranged on a thematic basis; structural earthquake engineering papers are followed by geotechnical earthquake engineering papers and those concerning tsunamis. In each category, overview papers are given priority over papers dealing with specific topics.

The contributing authors of the present book are all distinguished personalities of their respective areas expertise. The following paragraphs present brief biographical notes introducing the keynote speakers in the order their contributions appear in the volume.

M. A. Sozen – Following a long and successful academic career at the University of Illinois, Urbana, he has been teaching at Purdue University as the Kettelhut Distinguished Professor of Structural Engineering since 1993. His current research focuses on vulnerability assessment of building and transportation structures including effects of earthquake, impact, and explosion.

U. Ersoy – Following a long and successful academic career at the Middle East Technical University, he is currently teaching at Bogazici University. He has significantly contributed to the advancement of civil engineering in Turkey through publishing books, developing codes, promoting experimental research, advising engineers, besides teaching at the university.

J. Jirsa – Professor and Chair of the Department of Civil, Architectural and Environmental Engineering and Janet S. Cockrell Centennial Chair in Engineering, University of Texas at Austin. He was elected to the National Academy of Engineering in 1988. His research focuses on durability of reinforced concrete structures during earthquakes and seismic repair and strengthening of structures.

R. Spence – Professor of Architectural Engineering in the Department of Architecture at Cambridge University and a Fellow of Magdalene College. His principal research and consultancy interests are disaster risk assessment and disaster mitigation. He has directed numerous research and consultancy contracts and he is the author of several books and more than 150 technical papers. Clients for projects on risk assessment (earthquake, windstorm and subsidence) have included insurers, reinsurers and brokers and engineering consultants.

T. P. Tassios – Following a long and successful academic career at the National Technical University of Athens, he is still teaching at the same university as a Professor Emeritus. He made important contributions to the development of the model code of the European Concrete Committee out of which the Eurocode has evolved. His research interests include reinforced concrete and masonry structures under seismic conditions.

M. Dolce – Professor of Earthquake Engineering at the University of Naples Federico II and Director of Seismic Risk, Department of Civil Protection, Italy. He has also been consulting various public and private agencies. His research activity related to seismic behaviour of masonry and reinforced concrete buildings and the use of smart materials and passive control techniques, has resulted in four books and seven patents, besides over 300 papers.

K. Ishihara – Following a long and successful academic career at the University of Tokyo, he has been teaching at the Tokyo University of Science and then at Chuo University. He was an active member and one term President of the International Society for Soil Mechanics and Foundation Engineering. He is the recepient of various awards and author of the book "Soil Behaviour in Earthquake Geotechnics". His work has focused on modelling of cohesionless soils, dynamic pore pressure and liquefaction and seismic stability of foundations and dams.

M. Hamada – Professor in the Department of Civil and Environmental Engineering, Waseda University, Japan. He has been an active member and one term President of the Japan Society of Civil Engineers. With the goal of contributing to the creation of a safe and secure society against natural disasters, his research concentrates on the development of sustainable infrastructures for earthquake disaster mitigation.

A. Ansal – Professor in the Earthquake Engineering Department of Bogazici University. Presently, he is the Secretary General of European Association for Earthquake Engineering; Editor of Bulletin of Earthquake Engineering; Editor-in-Chief of the Book Series on "Geotechnical, Geological and

Earthquake Engineering" by Springer; President of the Turkish National Committee on Earthquake Engineering. His main areas of interest are microzonation methodologies, earthquake scenarios, effects of geotechnical factors on earthquake damage, cyclic behaviour of soils, liquefaction.

M. Saatcioglu – Professor of Structural Engineering and University Research Chair in the Department of Civil Engineering of the University of Ottawa. He is currently the President of the Canadian Association for Earthquake Engineering, Director of the Hazard Mitigation and Disaster Management Research Centre of the University of Ottawa and the Director of the Ottawa-Carleton Earthquake Engineering Research Centre. His research interests include design, analysis and retrofit of structures subjected to extreme loads, including those caused by earthquakes, tsunamis and bomb blasts.

W. Lanmin – Professor of Geotechnical Engineering and presently the Director of Lanzhou Institute of Seismology and the Director of the National Registered Geotechnical Engineers of China. He has authored four books and over 100 academic and technical papers. His research interests lie in soil dynamics and geotechnical earthquake engineering.

The present volume contains material distilled from vast experience of distinguished experts of earthquake and tsunami disaster mitigation, and it is being brought to the attention of practicing engineers, administrative authorities and academia with the hope that it may shed some light on their future research and planning activities and their everyday practice.

Ankara, Turkey A. Tugrul Tankut

Contents

Contributors

Atilla Ansal Kandilli Observatory and Earthquake Research Institute, Boğaziçi University, Istanbul, Turkey, ansal@boun.edu.tr

Mauro Dolce Civil Protection Department, General Director of Seismic Risk Office, Rome, Italy, mauro.dolce@protezionecivile.it

Uğur Ersoy Civil Engineering Department, Boğaziçi University, Istanbul, Turkey, ugur.ersoy@boun.edu.tr

Masanori Hamada Department of Civil and Environment Engineering, Waseda University, Tokyo, Japan, hamada@waseda.jp

Kenji Ishihara Research and Development Initiative, Chuo University, 1-5-7 Kameido, Koto-ku, Tokyo, 136-8577 Japan, ke-ishi@po.iijnet.or.jp

James O. Jirsa Department of CAEE, University of Texas at Austin, Austin, TX 78712, USA, jirsa@uts.cc.utexas.edu

Sun Junjie Lanzhou Institute of Seismology, China Earthquake Administration, 450 Donggangxilu Ave., Lanzhou 730000, China, sunnjunj@163.com and sunjj@gssb.gov.cn

InSung Kim Degenkolb Engineers, 235 Montgomery, Suite 500, San Francisco, CA 94104, USA, kim@degenkolb.com

Aslı Kurtuluş Kandilli Observatory and Earthquake Research Institute, Boğaziçi University, Istanbul, Turkey, asli.kurtulus@boun.edu.tr

Wang Lanmin Lanzhou Institute of Seismology, China Earthquake Administration, 450 Donggangxilu Ave., Lanzhou 730000, China, Wanglm@gssb.gov.cn

Murat Saatcioglu Department of Civil Engineering, University of Ottawa, Ottawa, Canada, murat.saatcioglu@uottawa.ca

Mete A. Sözen Purdue University, West Lafayette, Indiana, USA, sozen@purdue.edu

Robin Spence Cambridge Architectural Research Ltd., 25 Cwydir Street #6, Cambridge, CBI 2LG, UK, r.spence@carltd.com

T.P. Tassios National Technical Univiversity of Athens, Athens, Greece, tassiost@central.ntua.gr

Gökçe Tönük Kandilli Observatory and Earthquake Research Institute, Boğaziçi University, Istanbul, Turkey, gokce.tonuk@boun.edu.tr

Wu Zhijian Lanzhou Institute of Seismology, China Earthquake Administration, 450 Donggangxilu Ave., Lanzhou 730000, China, zhijianlz@163.com

Is The Domed City Doomed?

Mete A. Sözen

The earthquake risk that Istanbul, the City of Domes, sustains has been investigated by many experienced and knowledgeable scientists and engineers ad infinitum (examples are Griffiths et al. 2007 and Parsons, 2004). Realistically, there is no room left for another opinion. Accordingly, the following text has been written appropriately in the shadow of Mevlana, who advised that on such matters "Öyleyse Söz Kısa Olmalı, Vesselam!"[1], to assemble a few simple facts to ask the reader to draw his/her own conclusions.

Figure 1 shows the sequence of earthquakes that have caused serious damage in Istanbul from the fourth century on (Sakin 2002) and contains a linear trendline that has no basis in science. But it does raise the questions: Will there be a strong-motion event #15? Will it occur sometime between 2009 and 2050?

The cautious answer is that it may or it may not. The cautious response may be to plan that it will occur and rejoice if it does not. Investment to avoid the catastrophe before the event is likely to be much cheaper than the cost of losses in lives and property after the event, if it occurs.

Insofar as it is known, the strong-motion events occurring up to the 20th century may have imposed essentially similar demands on buildings in the City. The construction in the City has remained essentially the same mix and low-rise masonry except for the domed structures until mid-twentieth century. As of the 1950's, urban population has exploded from approximately a million to approach an alleged 15 million including the sprawling suburbs. This burst has been accompanied by a similar explosion in reinforced concrete construction. The quality of the construction has been meticulously captured by a series of studies made by Özcebe et al. (Ozcebe, 2006; Yakut, 2006). They have measured and archived the essential properties of some 12,000 buildings, an immensely valuable database.

M.A. Sözen (✉)
Purdue University, West Lafayette, Indiana, USA
e-mail: sozen@purdue.edu

[1] This brief sentence, loaded with meaning, denies translation but one can translate it freely as "in that case, cut it short!"

A.T. Tankut (ed.), *Earthquakes and Tsunamis,* Geotechnical, Geological, and Earthquake Engineering 11, DOI 10.1007/978-90-481-2399-5_1,

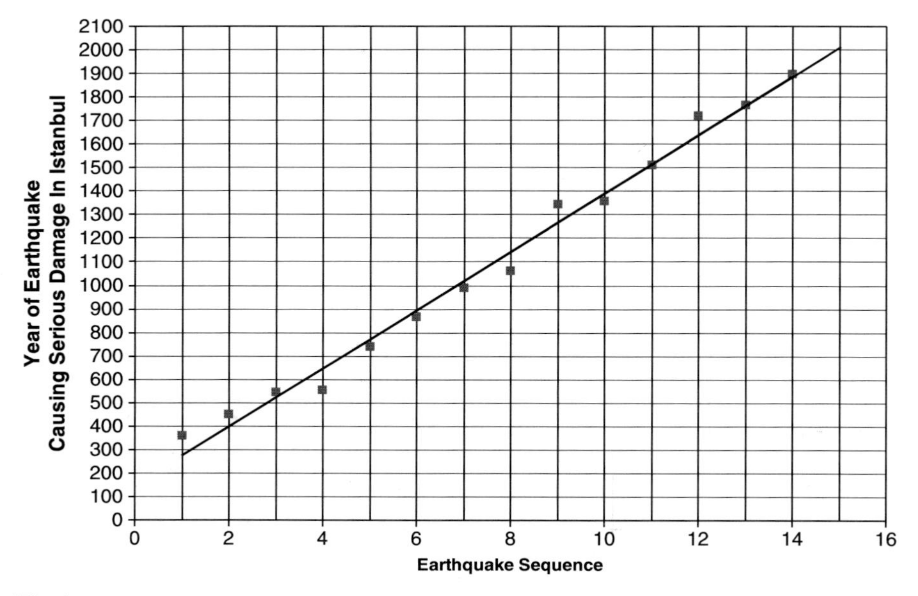

Fig. 1

The Hassan Index (Hassan, 1994) has been tested in Turkey on the basis of data from earthquakes of Erzincan, 1992, Marmara, 1999, Düzce, 1999, and Bingöl 2003 (http: //www.anatolianquake.org/). The index (Fig. 2), applicable to low-rise construction (up to approx. seven stories), is very simple and manifestly naïve. It is defined by two simple co-ordinates, a Column Index, CI, and a Wall Index, WI.

$$\mathrm{CI} = \frac{\Sigma \mathrm{A_c}}{2 \cdot \Sigma \mathrm{A_f}}$$

$$\mathrm{WI} = \frac{\Sigma \mathrm{A_w} - \Sigma \mathrm{A_m}}{\Sigma \mathrm{A_f}}$$

ΣA_c: Sum of the cross-sectional areas of columns at base
ΣA_w: Sum of cross-sectional areas of reinforced concrete walls at base in one direction
ΣA_m: Sum of cross-sectional areas of unreinforced masonry walls at base in one direction
ΣA_f: Sum of all floor areas above base

Clearly, the Hassan Index is insensitive to the intensity of the earthquake demand, the period of the building, the quality of the materials, the details of the reinforcement. And yet it has been shown to provide a reasonable estimate of the vulnerability of low-rise (seven stories and lower) reinforced concrete structures with respect to the data in four different strong-earthquake events. The overall conclusion from the

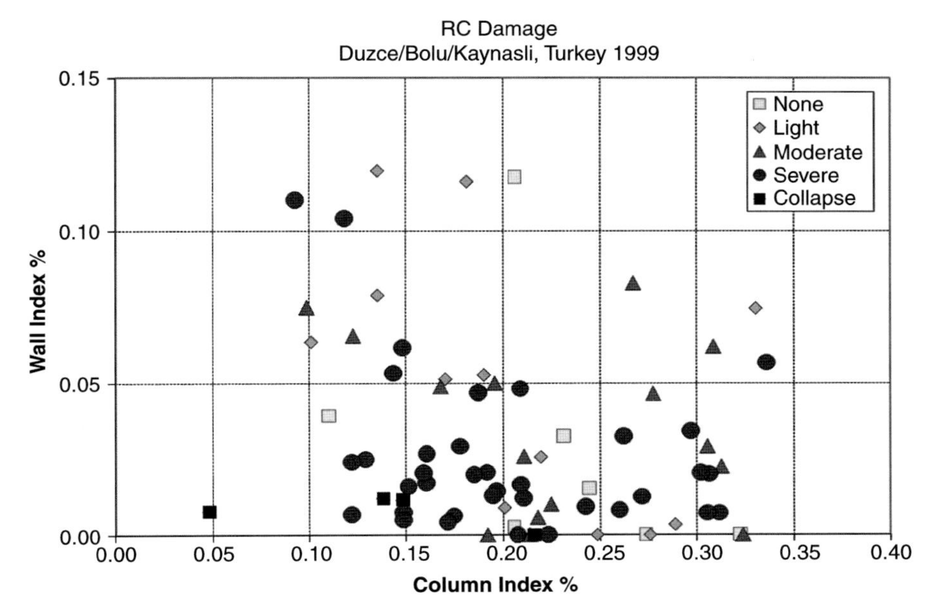

Fig. 2 (See also Color Plate 1 on page 209)

calibrations of the Hassan Index is that if a structure has a CI below 0.25% and a WI below 0.15%, it is in jeopardy.

Figures 3 and 4 show the CI indices for the building inventory archived by Ozcebe et al. Of the buildings inventoried in Küçükçekmece and Zeytinburnu, 90% have Hassan CI indices below 0.3%. Reinforced concrete walls were reported in a negligibly small fraction of the building inventory. The CI index is the governing one in all but a few instances.

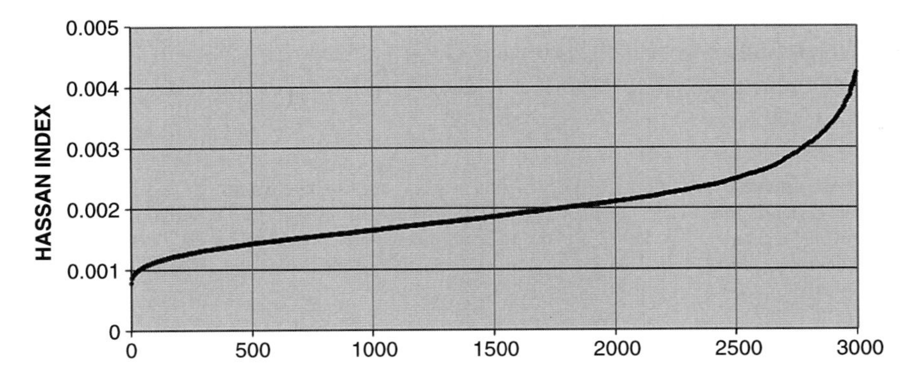

Fig. 3

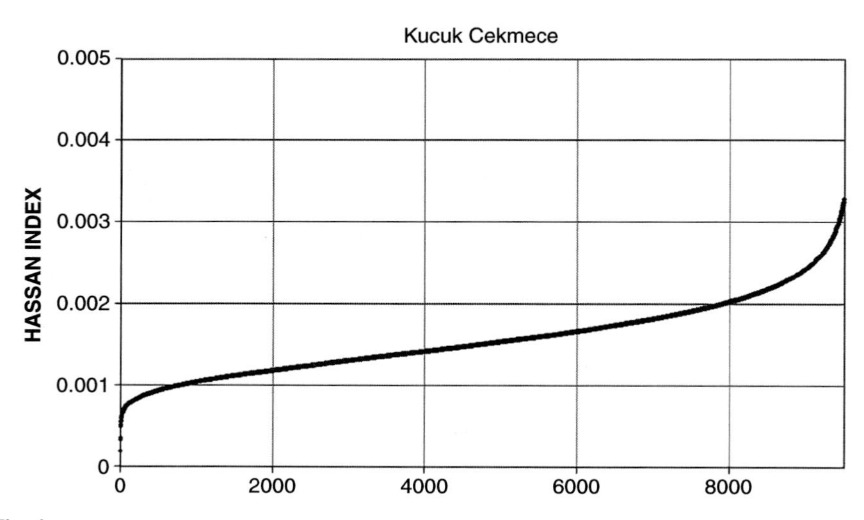

Fig. 4

Concluding Question

It is possible that the trendline in Fig. 1 is misleading. Possible damaging event #15 may not occur.

It is possible that the Ozcebe database is flawed.

It is possible that the Hassan index is wrong.

Given those possibilities, the following four combinations present themselves.

Combination 1: The trendline is wrong. Ozcebe is right. Hassan is right.
Combination 2: The trendline is right. Ozcebe is wrong. Hassan is wrong.
Combination 3: The trendline is right. Ozcebe is right. Hassan is wrong.
Combination 4: The trendline is right. Ozcebe is right. Hassan is right.

Any interested reader should be able to estimate, if approximately, the relative likelihoods of the four combinations listed or at least identify the one that is sensible for making plans. Readers who live in the Domed City should think about them seriously and make their plans accordingly.

Acknowledgments The work leading to this paper was initiated under sponsorship of U.S. National Foundation Grant CMS0512964. The writer is indebed to Dr. Güney Özcebe of Middle East Technical University for the building data, without precedent, on Zeytinburnu and Küçükçekmece.

References

Griffiths, J. H. P., A. Irfanoĝlu, and S. Pujol, "Istanbul at The Threshold: An Evaluation of the Seismic Risk in Istanbul," Earthquake Spectra, 23(1), February 2007.

Hassan, A. F. and M. A. Sözen, "Seismic VulnerabilityAssessment of Low-Rise Buildings in Regions with Infrequent Earthquakes," ACI Structural Journal, 94, 1994, 31–39.

Japan International Cooperation Agency (JICA), "The Study of Disaster Prevention/Mitigation Basic Plan in Istanbul including Seismic Microzonation in The Republic of Turkey: Final Report," Tokyo, Japan, 2002.

Ozcebe, G. et al., "In Defence of Zeytinburnu," NATO Science Series IV: Earth and Environmental Sciences; Advances in Earthquake Engineering for Urban Risk Reduction, 2006, 95–116.

Ozcebe, G. et al., "Seismic Risk Assessment of Existing Building Stock in Istanbul: A Pilot Application in Zeytinburnu District," Proc. 8th US National Conference on Earthquake Engineering, San Francisco, 2006.

Parsons, T., "Recalculated probability of M >= 7 Earthquakes beneath The Sea of Marmara, Turkey," Journal of Geophysical Research, 109, 2004.

Sakin, O., "Tarihsel Kaynaklariyla Istanbul Depremleri," Kitabevi 173, Istanbul, 2002(Turkish).

Yakut, A., G. Ozcebe, and M.S. Yucemen, "Seismic Vulnerability Assessment Using Regional Empirical Data," Earthquake Engineering and Structural Dynamics, 35, 2006, 1187–1202.

Seismic Rehabilitation of Reinforced Concrete Buildings

Uğur Ersoy

Abstract Turkey is located on one of the major earthquake belts of earth. In the past fifteen years five major earthquakes caused significant damages and loss of lives. Turkish government sponsored comprehensive repair and strengthening projects after each of these earthquakes.

In this paper, research in Turkey aimed to develop repair and strengthening techniques is summarized. The applications of these techniques to the buildings damaged during the earthquakes and to the existing vulnerable buildings are also discussed. Considering the characteristics of buildings and common deficiencies observed, system improvement seems to be the most feasible technique for Turkey. System improvement by introducing reinforced concrete infills to selected bays of framed structures is the most commonly used technique for rehabilitating damaged buildings. In this paper, experimental research and field applications related to this technique are discussed.

In the past eight years extensive research have been carried on in Turkey aimed to develop rehabilitation techniques to strengthen vulnerable buildings without evacuating the buildings. As a result of this research, techniques have been developed in which the nonstructural infill masonry walls are strengthened by either CFRP sheets or precast panels. Experiments have revealed that these strengthened infills behave almost like structural walls. In the paper the related research and applications are discussed.

In seismic rehabilitation, analyses are made to estimate the safety and performance of the building prior to and after rehabilitation. In modeling the structure for analyses the engineer is faced with numerous uncertainties. Therefore he or she has to make assumptions related to these uncertain parameters. In this paper problems involved in modeling the structure are briefly discussed.

U. Ersoy (✉)
Boğaziçi University, Civil Engineering Department, Istanbul, Turkey
e-mail: ugur.ersoy@boun.edu.tr

A.T. Tankut (ed.), *Earthquakes and Tsunamis,* Geotechnical, Geological, and Earthquake Engineering 11, DOI 10.1007/978-90-481-2399-5_2,

1 Introduction

Structural repair and structural strengthening are not new concepts. One can easily say that structural rehabilitation should be as old as the history of building engineering. However, scientific approach to seismic rehabilitation probably dates to 20th century. Research on seismic rehabilitation accelerated in the second half of 20th century. The development of rehabilitation techniques largely depended on experimental research.

In Turkey, the first experimental research on seismic rehabilitation was initiated in 1968. A group from Middle East Technical University (METU) strengthened a one story framed building by filling selected bays of the structure with reinforced concrete walls after the 1968 Bartin earthquake. One story reinforced concrete unfilled frames were tested at METU to understand the behavior of such structures. Since then, the experimental research at METU has been focused on the behavior of repaired and/or strengthened reinforced concrete structures (Ersoy, 1996).

The problem of seismic rehabilitation can be classified into two categories; (a) repair and strengthening of structures damaged during earthquakes and (b) strengthening of existing structures which are found to be vulnerable. The technique or method of rehabilitation to be used depends on the nature of the problem, architectural constraints and functional requirements. For example, in choosing the rehabilitation technique for an industrial building, the most important factor is how soon the plant can be put into operation. Usually the cost of a few days of delay in production is higher than the total cost of the building!

In structural rehabilitation, evaluation of seismic performance of damaged or undamaged existing building is very important. The evaluation is usually made at several stages, starting with fast screening. Evaluation is outside the scope of this paper. However it should be pointed out that evaluation or assessment is the first and the crucial step in seismic rehabilitation. It can be compared with medical diagnosis made prior to the treatment of the patient!

In Turkey, comprehensive rehabilitation projects have been carried on after the major earthquakes in the past 15 years. Hundreds of buildings having medium damage have been repaired and strengthened using the techniques developed as a result of research made in Turkey. Extensive surveys have been carried on to assess the damaged and undamaged buildings. Data obtained from such surveys have and will be used in developing rehabilitation strategies.

2 Strategy for Seismic Rehabilitation

In Turkey and in some other countries located in seismic zones a large portion of existing buildings have non-ductile structural systems. Such structures demonstrated unsatisfactory performances during the past earthquakes. Many of these buildings either collapsed or were severely damaged.

It is not feasible to demolish and replace all the structures which are found to be seismically vulnerable. The strategy should be to retrofit these structures and demolish the ones for which retrofitting is not feasible. However considering the huge stock of vulnerable buildings and funds available, one can easily conclude that to accomplish such a task will take years. Then it will not be wrong to assume that we will have to deal with many damaged buildings after each near future earthquake. For this reason, repair and strengthening of damaged buildings should be included in the "seismic rehabilitation strategy."

The available methods and technologies for seismic retrofitting of damaged and undamaged reinforced concrete buildings can be classified into two groups; (a) member retrofitting and (b) system improvement. Although member retrofitting is known to most of the engineers, "system improvement" needs to be defined and discussed (Ersoy and Tankut, 1992).

In system improvement, the inadequate lateral load resisting system (usually framed structures) of the building is replaced by a new system which is composed of rigid vertical members. In such applications the existing load carrying system is assumed to contribute only to gravity loads and imposed deformations. The basic idea in system improvement is to convert selected bays of the structure in each direction into structural walls. Most commonly used system improvement techniques are given below:

a – Reinforced concrete infilled frames
b – Precast reinforced concrete infilled frames
c – Steel cross bracing
d – Strengthening non-structural infills by carbon fiber (CFRP) layers or by precast panels.

The most commonly used system improvement method used in Turkey is reinforced concrete infilled frame technique. Steel bracing is not recommended unless special details are developed, because in such systems large forces are applied to beam-column joints which are deficient in Turkey.

Reinforced concrete infilled frame technique has been investigated in Turkey by extensive research since 1968. It was applied to hundreds of damaged buildings after the Bartin (1968), Erzincan (1992), Dinar (1995), Ceyhan (1998) and Marmara and Düzce (1999) earthquakes.

Although reinforced concrete infilled frame technique has proved to be safe, economical and practical, it requires the evacuation of the building prior to application. This is not a handicap for structures damaged during the earthquake. However, when strengthening of large number of existing, undamaged buildings are considered, it will not be practical to evacuate the buildings. In recent years techniques for strengthening have been developed which will not require evacuation of buildings. Such techniques are listed as (d) in the above list.

System improvement becomes feasible when one or more of the following conditions exist.

a – When the lateral stiffness of the frame system is inadequate.
b – When too many members have to be repaired and/or strengthened.
c – When the frame system has inherited weaknesses such as soft story, short column etc.

The seismic rehabilitation method or technique to be used should be chosen considering the problems involved, architectural restrictions, functional requirements and economy.

It will be helpful to clarify what is meant by problems involved. The structure might have lateral stiffness problem or some members might have deficiencies in moment capacity, shear capacity or axial load capacity. In some cases inadequate ductility might be the main problem. Therefore, first the aim or objective should be clearly defined and then the technique of rehabilitation should be chosen. For example if the main objective is to upgrade the moment capacity of the column, steel jacketing should be overruled since moment capacity cannot be improved with such a technique.

It is also very important to know whether the building can be evacuated or not during the rehabilitation operations. The rehabilitation technique to be used depends very much on this issue.

3 Member Rehabilitation

After assessing the present condition of the building, the engineer may conclude that the structure can reach the aimed safety and performance level by rehabilitating some of the structural members, such as columns, beams and structural walls.

Different techniques are used in the repair/strengthening of reinforced concrete members. The most commonly used ones are given below:

- Column or beam jacketing (reinforced concrete or steel jacket)
- Adding a new reinforced concrete layer to the member.
- Fastening steel plates to the member
- Fastening carbon or glass fiber reinforced polymer sheets to the member
- Wrapping the member with carbon fiber reinforced polymer sheets (CFRP).

Member rehabilitation is made to improve; (a) flexural strength, (b) shear strength, (c) axial load capacity, (d) confinement or (e) stiffness. Prior to choosing the rehabilitation technique to be used, the engineer should define the objectives clearly.

What properties of the member does he or she want to improve? In the following paragraphs the function of above listed techniques will be summarized (Ersoy and Tankut, 1992).

- Reinforced concrete column jacketing. It can increase the axial load and moment capacity of the column. To increase the moment capacity, column longitudinal bars have to be continuous thorough the floor. Column jacketing can also improve the ductility provided that the unsupported length of ties in the jacket is shortened by intermediate anchors. Ties provided in the jacket also increase the shear strength

- Steel jacketing for columns. Steel jacketing by placing steel angles at each corner of the column which are held together by welded transverse steel straps increases the axial load and the shear capacity. If the transverse straps are closely spaced and are in contact with the faces of the column through non-shrink mortar, then the confinement can also be improved. Steel jacketing should not used for increasing the flexural strength of the column.
- Adding a new reinforced concrete layer. This technique is used to increase the flexural strength of beams and slabs. The reinforcing bars in this added layer should be property anchored.
- Fastening steel plates to members. Steel plates anchored to the member are used to increase the flexural strength of beams.
- Fastening fiber reinforced polymers to members. Usually carbon fiber and glass fiber reinforced polymers are used for increasing the flexural capacity of beams at critical regions. These layers should be attached to the member by special anchors.
- Wrapping the member with fiber reinforced polymer layers. Wrapping of beam and columns with carbon or glass fiber reinforced polymers will increase the shear capacity. Confinement can be provided if the unsupported length of polymer layers is shortened by intermediate anchors.

The engineer involved in seismic rehabilitation should be familiar with the related research and should know the behavior of rehabilitated members. Problems related to lapped splices in column longitudinal bars will be taken as an example. The behavior of a column rehabilitated by jacketing will be affected adversely by the presence of lapped splices in column longitudinal bars made at floor level with inadequate lap length. Various techniques have been proposed to over-come the deficiency created by lapped splices. One of these techniques is to wrap the splice region with carbon fiber reinforced polymer sheets (CFRP). Wrapping with CFRP will provide confinement which will create a positive effect on bond. However it should be noted that this is true only if deformed bars have been used as column longitudinal reinforcement in which splitting stresses are dominant. Confinement will not significantly improve the behavior if plain bars have been used as column longitudinal reinforcement.

Another example is for columns repaired/strengthened under existing axial load. In general column jacketing is made under load. Tests have revealed that if a damaged column is repaired/strengthened under load, only the jacket concrete and reinforcement should be taken into consideration in calculating the stiffness and the strength of the column section. If the jacketing is applied to an undamaged column, the jacketed section and the existing section can be taken into account.

4 System Improvement

System improvement was defined and different techniques used were discussed in the previous section of this paper. In this section only two of these techniques will be discussed; (a) reinforced concrete infilled frames and (b) frames with strengthened non-structural infills (by CFRP layers and precast panels).

4.1 Reinforced Concrete Infilled Frames

First application of infilled frame technique in Turkey dates back to 1968. Experimental research on this subject was initiated immediately after this application. In the following thirty years several major research projects were carried on at the Middle East Technical University (METU) Structural Mechanics Laboratory. The first two projects were aimed to understand the behavior of infilled frames and to investigate the influence of different variables and connections. At the end of these test programs the following conclusions were reached (Ersoy, 1996; Altin et al., 1922).

- The infilled frame behaves as a structural wall with boundary columns if the infill is properly connected to the frame members. It was found out that connections provided by dowels bonded into the drilled holes by epoxy resulted in satisfactory performance.
- Strength and stiffness of boundary columns have a significant effect on the behavior of the infilled frame. In Fig. 1, curve no.1 is the lateral load – displacement envelope curve for the reference specimen (monolithic) and curve no.2 is envelop curve for the infilled frame (Altin et al., 1922). On the same

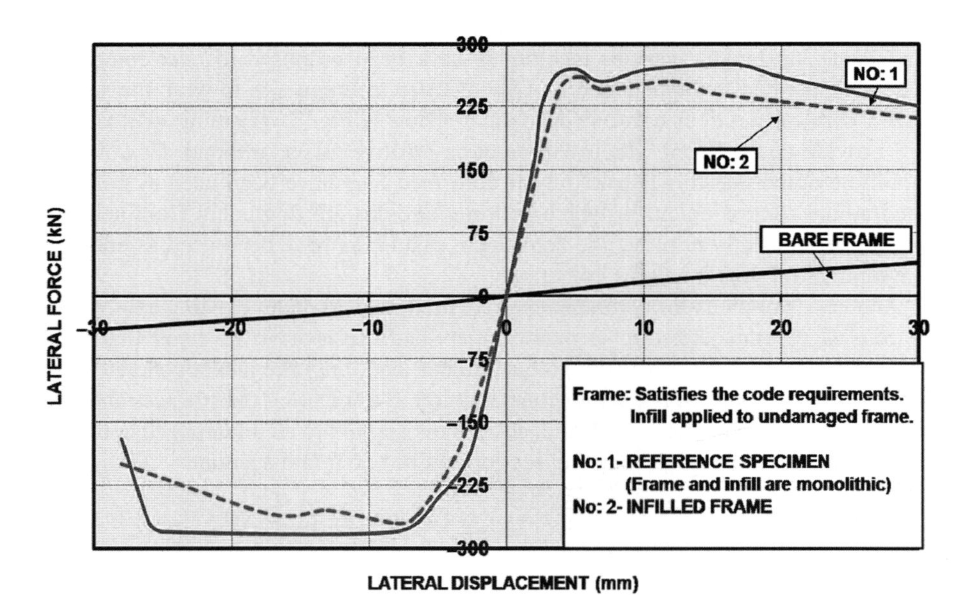

Fig. 1 Envelop curves of structural wall, reinforced concrete infilled frame and the bare frame

figure the envelope curve of the bare frame is also shown. Slope of the curves shown in this figure is a measure of stiffness. When the envelop curves of the bare frame and the infilled frame are compared, the increase in both the strength and stiffness becomes obvious. When curves 1 and 2 are compared, it is observed that the infilled frame behaved almost as good as the monolithic reference specimen.

In the first two projects, the infills were applied to undamaged frames. In order to keep the variables at a minimum level, the frames in these tests were designed and detailed in accordance with the Turkish Seismic Code (TSC). However in practice this technique (infilled frames) is used in the repair and strengthening of damaged frames. A great majority of these frames were poorly detailed using plain bars and had very low concrete strength. Therefore in later research projects the test specimens were designed and detailed poorly to simulate these buildings. The test specimens had the following deficiencies which were commonly observed in existing buildings.

- Low concrete strength (about 10 MPa)
- Inadequate confinement at the ends of columns and beams.
- Plain bars were used to reinforce the test frames.
- Ties had 90° hooks at each end.
- No ties were provided at beam – column joints.
- Lapped splices in column longitudinal bars were made at floor level. The lap length was inadequate (12–40 bar diameters).

To make the tests more realistic, the infills were introduced to damaged frames having the deficiencies listed above (Sonuvar et al., 2004; Türk et al., 2006).

Although the behavior and strength of these specimens were not as good as those of specimens in the first two test programs, still there was a significant improvement in behavior as compared to that of a bare frame.

In Fig. 2 lateral load – top story displacement envelope curves of one of the test specimens in given together with that of the bare frame. As can be seen, although the infill was introduced to a damaged, deficient frame (lap splice length: 40Ø), the strength and behavior as compared to the bare frame was significantly improved (Sonuvar et al., 2004).

In this specimen the length of lapped splices in column longitudinal bars was 40 bar diameters. In some other specimens the lap length was reduced to twenty bar diameters.

Tests have revealed that among the deficiencies listed previously, the one which affects the infilled frames behavior most adversely is the lapped splice with inadequate length (Sonuvar et al., 2004; Valluvan et al., 1993). The presence of such lapped splices in column longitudinal bars reduced both strength and stiffness of the infilled frame.

To minimize this adverse effect in a practical way, it was decided to place four large diameter longitudinal bars at the boundaries of the infilled, confined by ties.

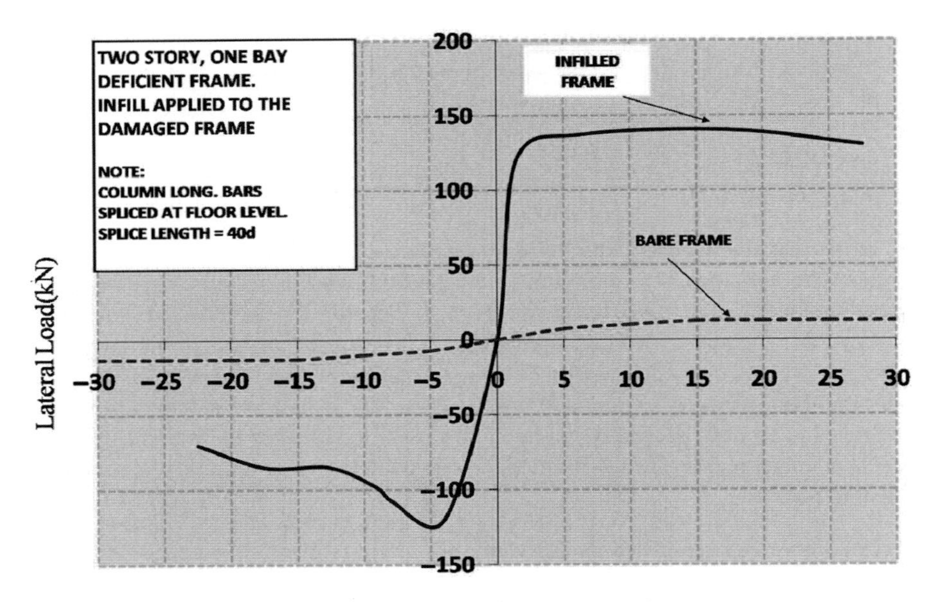

Fig. 2 Envelope curves

The continuity of the these bars through the floors was provided by dowels connected to the beams. Tests performed showed that with such additional bars at the boundaries, as shown in Fig. 3, the behavior was significantly improved (Sonuvar et al., 2004).

The specimens tested in these projects were one-bay, two-story, 1/3 scaled frames. In a later series, two-story, three bay frames were tested. After damaging the bare frame under reversed cyclic loading, it was strengthened by applying reinforced

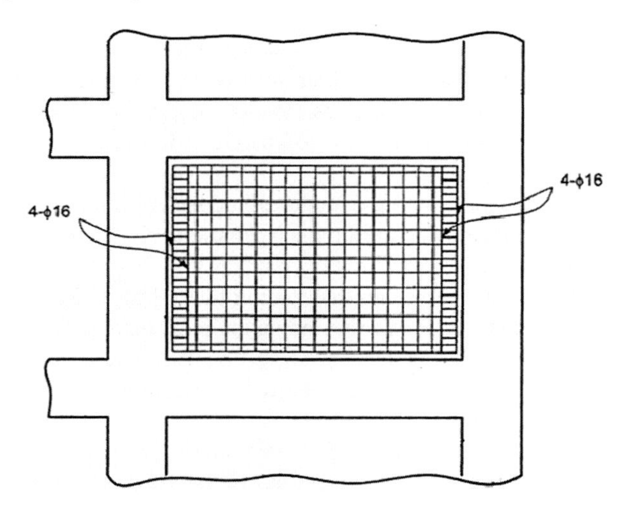

Fig. 3 Infilled frame with additional reinforcement at infill boundaries

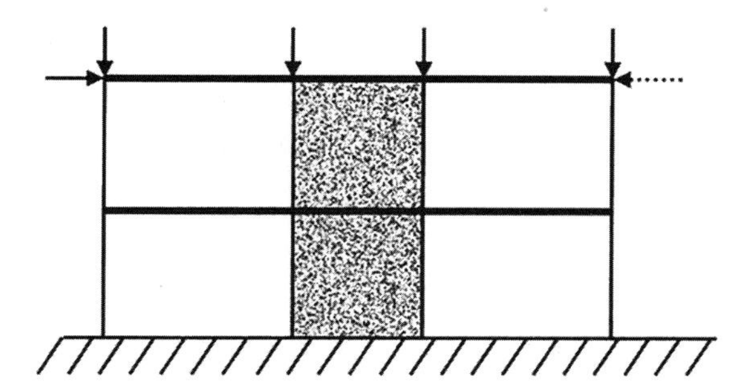

Fig. 4 Three bay test specimen

concrete infill only to the interior bay, Fig. 4. Although the damaged frame members were not repaired, the system behaved very well since a great portion of the lateral load was resisted by the infilled frame (Canbay et al., 2003).

The reinforced infilled frame technique was applied to hundreds of buildings after the 1968 Bartin, 1992 Erzincan, 1995 Dinar, 1998 Ceyhan and Marmara/Düzce earthquakes. As a result of these applications, it was concluded that the infilled frame technique is a feasible approach for the repair and strengthening of damaged buildings.

Based on laboratory tests, analytical studies and problems encountered in applications, METU developed recommendations to be used in the design and construction of reinforced concrete infilled frames.

4.2 System Improvement by Strengthening the Existing Non-Structural Infills

Although the reinforced concrete infilled frame technique proved to be a safe, economical and practical for the repair and strengthening of damaged buildings, it would not be feasible for the strengthening large number of existing buildings. The reason is that this technique can not be applied without the evacuation of buildings at least for three to five months.

In recent years, rehabilitation techniques were developed which would not require the evacuation of buildings. The basic idea in such techniques is to strengthen the existing non-structural infills which are expected to behave somewhat like structural walls.

Two infill strengthening techniques were investigated as a part of a comprehensive research project initiated by METU, financed by NATO and TÜBİTAK. Other contributors were; University of Texas (USA), Aristotle and Patras universities (Greece), St. Cyrill and Methodius University (Republic of Macedonia), Istanbul Technical University (Turkey), Kocaeli University (Turkey) and Boğaziçi University (Turkey).

In one of the infill strengthening techniques, Carbon Fiber Reinforced polymer (CFRP) sheets were used. After testing one-bay two-story frames strengthened by different patterns of CFRP, it was concluded that arranging CFRP sheets in a cross-bracing pattern as shown in Fig. 5 would be feasible (Özcebe et al., 2003).

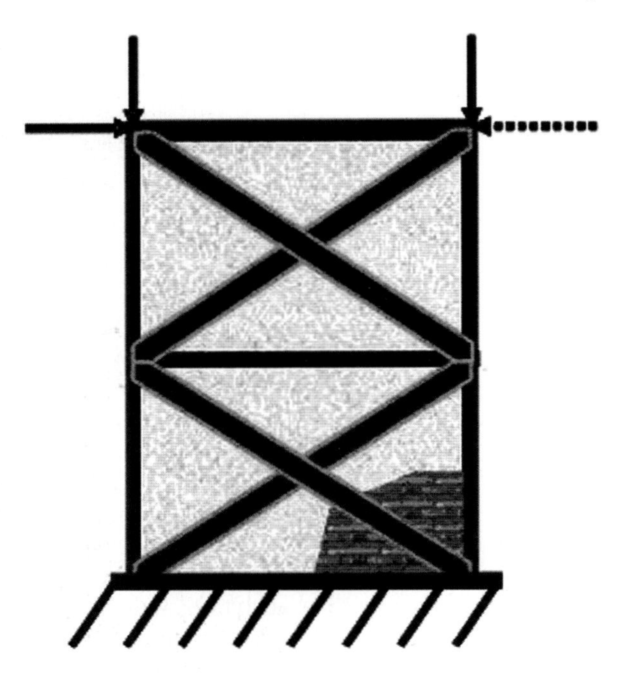

Fig. 5 Infill strengthening by CFRP

Special anchors were developed by rolling CFRP sheets. Using such anchors, CFRP sheets were connected to the infill and to the frame members. In Fig. 6, lateral load-drift ratio envelop curves for bare frame, reinforced concrete infilled frame (S1) and CFRP strengthened (S2) specimens are shown. CRFP strengthened specimen

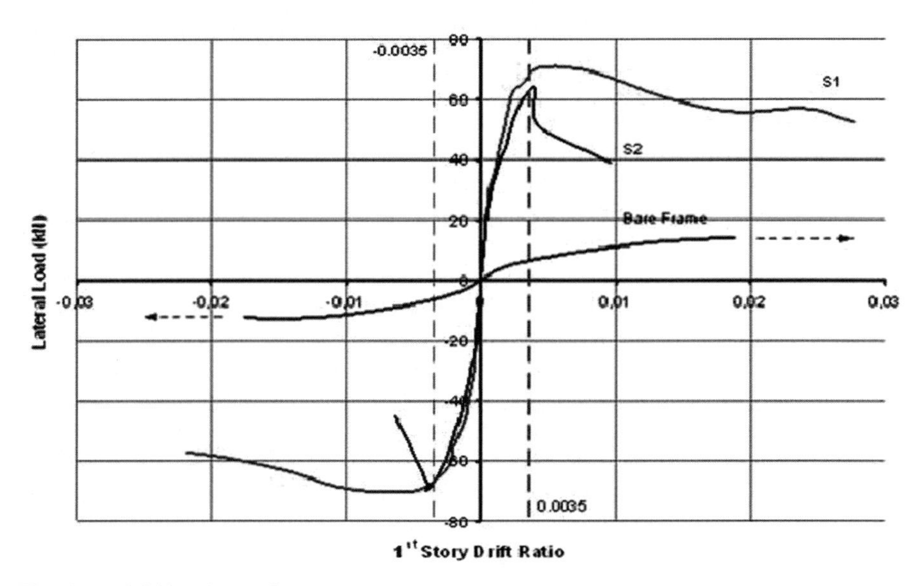

Fig. 6 Load-drift ratio envelop curves

Fig. 7 Rehabilitated building

S2 behaved almost like the reinforced concrete infilled frame (S1) up to the peak. However the failure of S2 was much more brittle as compared to S1. The brittle failure was due to failure of anchors (Özcebe et al., 2003).

Recently a four story reinforced concrete building in Antakya, Turkey was rehabilitated by strengthening the brick infills with CFRP sheets, Figs. 7 and 8. The rehabilitated structure was a residential building. After rehabilitation was completed, the occupants were interviewed. The interviewed occupants stated that the disturbance given was not more than what they experienced when the house was being painted. The contractor stated that he did not face any problems during application.

The second technique used for strengthening the infills is placing precast reinforced concrete panels on each face of the infill. The two type of panels used are shown in Fig. 9. At the end of the test program it was concluded that brick infilled frames strengthened by this technique behaved almost as good as the reinforced concrete infilled frames, provided that infills are properly connected to the frame members (Tankut et al., 2003). When compared with the bare frame, the ratios of

Fig. 8 CFRP Applications

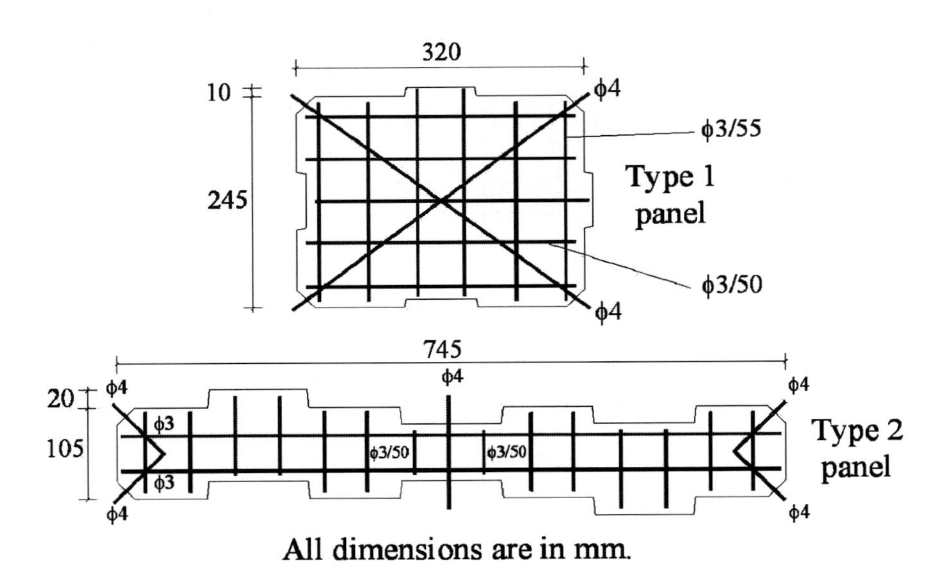

Fig. 9 Precast panels

strength and stiffness were 15 and 20 respectively. Compared to reference frame having unstrengthened brick infill, strength and stiffness ratios were 2.5 and 3.0 respectively.

5 Remarks on Rehabilitation

Rehabilitation of damaged buildings and seismic strengthening of existing structures have become a challenging task for the structural engineers. Large scale seismic rehabilitation projects have been carried on in Turkey and in other countries after the major earthquakes. Seismic strengthening of existing vulnerable structures is one of the major concerns in countries located in seismic regions.

Seismic rehabilitation of damaged and undamaged buildings is a very complex process, full of uncertainties. The engineer has to make analyses to evaluate the performance of the structure prior to and after rehabilitation. In modeling the structure for analyses, the engineer is faced with many unknowns and uncertainties which force him/her to make numerous assumptions. Therefore numbers which come out as a result of analyses are by no means exact. The engineer, who is aware of the uncertainties and the debatable nature of the assumptions made, does not consider these numbers as the solution of the problem but uses them as a guidance for his/her judgment and decision.

Analyses of existing structures (damaged or undamaged) is much more complex than designing a new building. In the design of a new building, the engineer starts with a clean page. However in retrofitting, the page is not clean. It is full of existing mistakes, damages, deficiencies unknowns and uncertainties. Therefore the engineer has to rely not only on analyses but also on his/her intuition, judgment and experience. Rehabilitation is not only science but art.

Retrofitting should not be attempted before understanding the problems involved in that particular building. If the building is damaged, the cause of damage should be clearly identified prior to retrofitting.

The choice of retrofitting technique depends very much on the nature of problems involved, architectural restrictions and functional requirements. It is also important to know the cost of interrupting the functionality. In some cases cost of interruption can be the determining factor in selecting the retrofit technique to be used.

Seismic rehabilitation requires a sound knowledge of seismic behavior. Due to the uncertainties in analyses, the engineer has to rely on his judgment, which is not possible without a good background on seismic behavior. Seismic rehabilitation made with limited knowledge and experience may not improve the performance, on the contrary in some cases it can make the structure more vulnerable.

There are no so called standard techniques in seismic rehabilitation. The methods and techniques used for some buildings should not be considered to be remedies for a different building. These techniques should be revised or new techniques should be developed considering the problems involved in that particular building.

In the light of the above discussion, it can be concluded that code and specifications for seismic rehabilitation should not go into the details but should provide a guideline to the engineer.

References

Altın, S., Ersoy, U., and Tankut, T., "Hysteretic Response of RC Infilled Frames", ASCE Structural Journal, 118(8), August 1992.

Canbay, E., Ersoy, U., and Özcebe, G., "Contribution of RC Infills to the Seismic Behavior of Structural Systems", ACI Structural Journal, 100(5), Sept.–Oct. 2003.

Ersoy, U., "Seismic Rehabilitation-Application, Research and Current Needs", 11. World Conference on Earthquake Engineering Proceedings, Mexico, 1996 (invited paper).

Ersoy, U., and Tankut, T., "Current Research at METU on Repair and Strengthening of RC Structures", Bulletin of ITU, 45(1–3), 1992.

Özcebe, G., Ersoy, U., Akyüz, U., and Erduran, I., "Rehabilitation of Existing RC Structures Using CFRP Fabrics", 13. World Conference on Earthquake Engineering, Vancouver, Canada, August 2003, Paper No: 1939.

Sonuvar, M.O., Özcebe, G., and Ersoy, U., "Rehabilitation of RC Frames with RC Infills", ACI Structural Journal, 100(4), July–August 2004.

Tankut, T., Ersoy, U., Özcebe, G., Baran, M., and Okuyucu, D., "In Service Strengthening of RC Framed Buildings", NATO Sfp Workshop on Seismic Assessment and Rehabilitation of Existing Buildings, NATO Science Series. 29, Kluwer Academic Publishers, Oct. 2003.

Türk, M., Ersoy, U., and Özcebe, G., "Effect of Introducing RC Infills on Seismic Performance of Damaged RC Frames", Structural Engineering and Mechanics, 23(5), 2006.

Valluvan, R., Kreger, M.E., and Jirsa, J.O., "Strengthening of Column Splices for Seismic Retrofit of Nonductile RC Frames", ACI Structural Journal, 90(4), July–August 1993.

Use of CFRP to Strengthen Splices and Provide Continuity in Reinforced Concrete Structures

James O. Jirsa and InSung Kim

Abstract Strengthening techniques using CFRP materials have been investigated for improving the capacity of lapped splices and to provide continuity of reinforcement in existing reinforced concrete structures in seismic regions. CFRP offers advantages because the strengthened elements do not change shape or appearance. The difficulties in using CFRP are associated with the limited force transfer that can be developed between the CFRP and the concrete element through adhesion. CFRP anchors transfer forces from CFRP sheets mounted on the concrete surface to the core of the concrete element. Anchors have been used to strengthen spices in rectangular elements or to strengthen beams at column supports where moment reversals cannot be developed because reinforcement details in codes at the time of construction did not require continuity of both top and bottom reinforcement.

1 Background

One of the main problems in rehabilitating reinforced concrete structures in seismic zones is the development of procedures that are inexpensive and do not substantially change the appearance or the usage of a structure. For frame elements, two problems that are common in structures that were designed and built prior to the late 1960's are the lack of continuity of bottom beam reinforcement through beam-column connections and column splices that were designed to carry compression only. In many cases, the structures were designed to carry only vertical loads so that positive beam moments at a joint were not likely and columns carried only compressive loads. However, moment reversals will occur in the beams and the columns and the longitudinal reinforcement may reach yield during an earthquake.

In previous studies, these deficiencies were addressed using steel jackets to strengthen splices and steel face plates were added to provide positive moment

J.O. Jirsa (✉)
Department of CAEE, University of Texas at Austin, Austin, TX 78712, USA
e-mail: jirsa@uts.cc.utexas.edu

A.T. Tankut (ed.), *Earthquakes and Tsunamis,* Geotechnical, Geological, and Earthquake Engineering 11, DOI 10.1007/978-90-481-2399-5_3,

Fig. 1 Strengthening using steel plates and jackets

capacity at the column (Aboutaha et al., 1999; Estrada, 1990). The use of steel jackets or steel plates for these situations is shown in Fig. 1.

The objective of this study was to determine if CFRP anchors could be used to improve the force transfer between a CFRP sheet and the concrete element so that the full strength of the sheet could be utilized. A CFRP anchor is shown in Fig. 2.

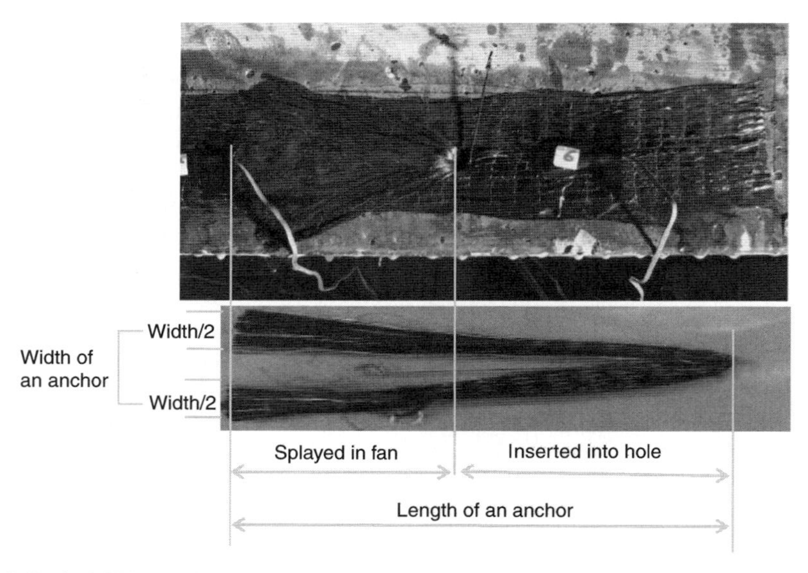

Fig. 2 Typical CFRP anchor

2 Experimental Study

Two series of tests were conducted:

- Impact loading (high strain rate) on beams with discontinuous flexural reinforcement.
- Cyclic loading of columns with inadequate splice lengths.

Studies of rehabilitation techniques using different geometries and quantities of CFRP materials for specimens subjected to static loading preceded the work reported here (Kim, 2006; Kim et al., 2008; Kobayashi et al., 2001; Orton et al., 2006; Ozdemir and Akyuz, 2006). The CFRP sheets in some of those tests developed ultimate tensile strength under static loading.

2.1 Impact Tests on Beams with Discontinuous Reinforcement

2.1.1 Test Specimens

Two different beam conditions were tested as shown in Fig. 3. The beam dimensions were 200 × 400 × 4880 mm (Fig. 3a) or 150 × 300 × 4880 m (Fig. 3b). Top reinforcement consisted of two 10 mm bars (Grade 420) and bottom reinforcement was two 19 mm bars with a discontinuity at the middle of the specimen. Sufficient shear reinforcement was provided to prevent shear failure. Five specimens were tested representing a situation where a flat CFRP sheet crossed the discontinuity (Fig. 3a) and two were tested with the discontinuity in the column (Fig. 3a). In the second case, the CFRP sheet was made continuous though the column by using a CFRP sheet rolled to produce a tendon through a hole drilled in the column (Fig. 3b). The measured ultimate tensile strength (1000 MPa) of the CFRP sheets was used in calculating the static strength of the test beams. The previous studies showed that

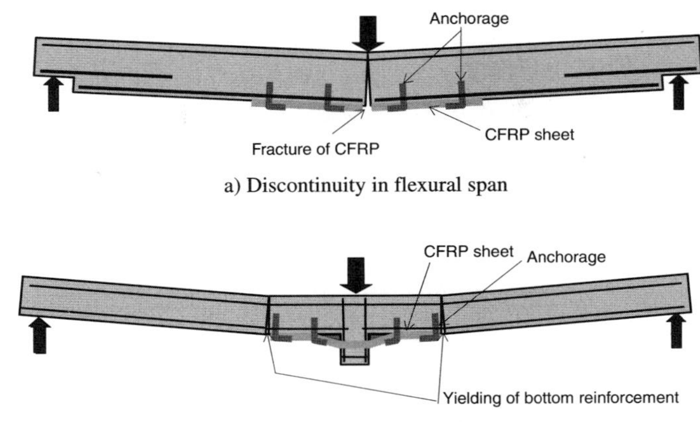

Fig. 3 Types of beam specimens

Table 1 Beam test details

Test	CFRP Repair	Layout of CFRP Repair	Conc., Mpa	Surface Prep.
Bm 1	No anchors		35	Sand blast
Bm 2	Anchors		14	Sand blast
Bm 3	Anchors	Same as Bm 2	35	Sand blast
Bm 4	Anchors	Same as Bm 2	14	No adhesion
Bm 5	Anchors	30% more CFRP than in Bm 2	35	No adhesion
Bm – Col 1	Anchors		42	Grinding
Bm-Col 2	Anchors		42	Grinding

the measured properties of this CFRP material were consistent with the specified properties from the manufacturer (Kim, 2006; Orton et al., 2006). Details of the test specimens are shown in Table 1 and all details can be found in (Kim, 2006).

For Beams 1–5, a CFRP sheet 152 mm wide × 1220 mm long was attached (Fig. 4). CFRP anchors (as shown in Fig. 2) were made using a sheet width of 50 mm per anchor except for Beam 5 where the width was increased by 30%. The length of the anchor was 230 mm with 140 mm of the anchor inserted in to a 10 mm-diameter-hole drilled into concrete, and the rest of the anchor was spread out in a fan shape on the CFRP sheet. Twelve anchors were installed as shown in Fig. 4.

The CFRP was attached on two different concrete surface conditions. For Beams 2 and 3, the concrete surface was sand-blasted before installing the CFRP materials.

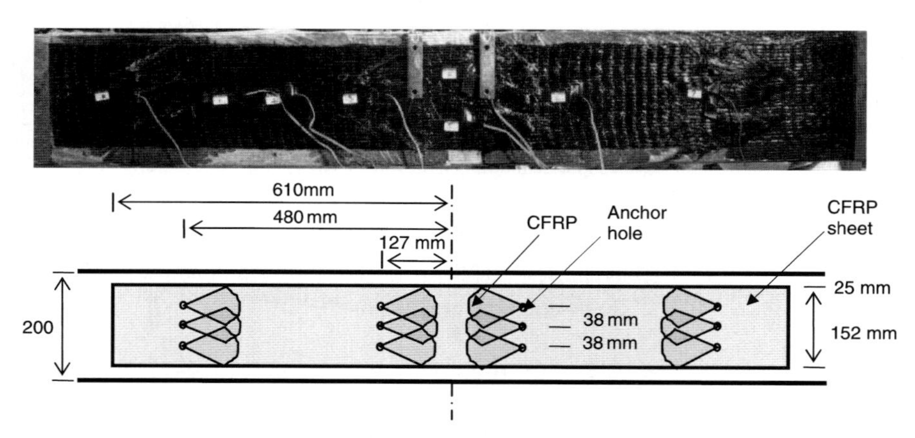

Fig. 4 Rehabilitation using CFRP on bottom face of Beams 2–5

Fig. 5 Surface condition before attachment of CFRP sheets

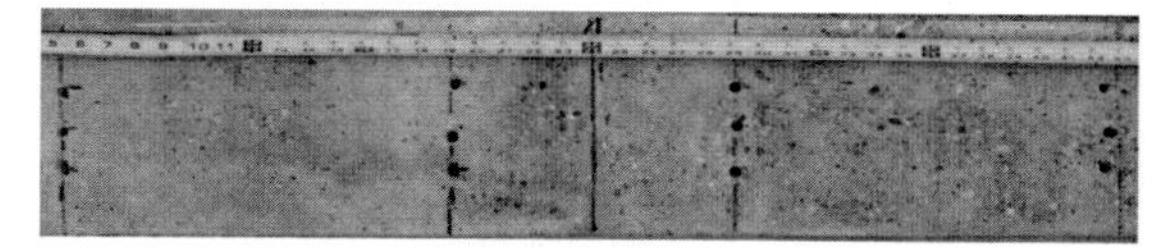

a) Sand-blasted concrete surface, Beams 2 and 3

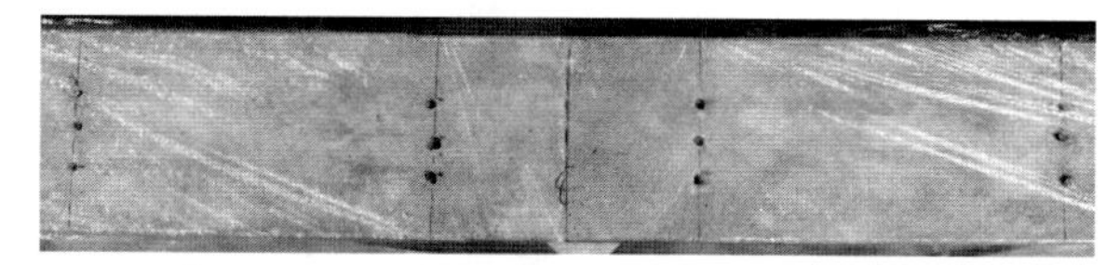

b) Separation using clear polyethylene wrap, Beams 4 and 5

For Beams 4 and 5, clear polyethylene wrap was placed on the surface to eliminate bond between the CFRP sheet and concrete. Forces in the CFRP sheet were transferred to the concrete only by the CFRP anchors in this case. In Fig. 5, the surface condition as well as the location of the anchor holes is shown.

2.1.2 Test setup

The overall test setup is shown below in Fig. 6. An impact load was applied to the middle of the beam with a 930 kg pendulum mass. The beam was placed on its side for testing. Drop heights of the pendulum mass were varied with respect to capacity of specimens. Load cells were installed in front of the pendulum mass (890 kN-capacity load cell) and at both supports (445 kN-capacity load cell) to measure applied load and reactions. Deflection at the middle of the beam was measured with two linear motion transducers. Strain gages were installed on the CFRP materials to measure development of tensile strains along the CFRP sheets.

2.1.3 Test Results and Failure Modes

The results of the tests with impact loading are summarized in Table 2. It is important to note that the modes of failure were consistent with previous tests where static loads were applied (Orton, 2007).

The different modes of failure are shown in Figs. 7, 8, 9 and 10.

From the tests the following comparisons can be made:

- Anchors worked equally well in concrete with 14 or 35 MPa strength
- Surface preparation was not critical, however, with no adhesion between the concrete and the CFRP sheet, it was necessary to increase the amount of material in the anchors by 30% to produce fracture of the sheet
- A combination of CFRP sheets and anchors enabled the beam to develop positive moment at the beam face where reinforcement discontinuity would prevent any positive moment development.
- By changing the length of the CFRP sheet, it was possible to create a condition where positive beam hinging occurred before the CFRP fractured.

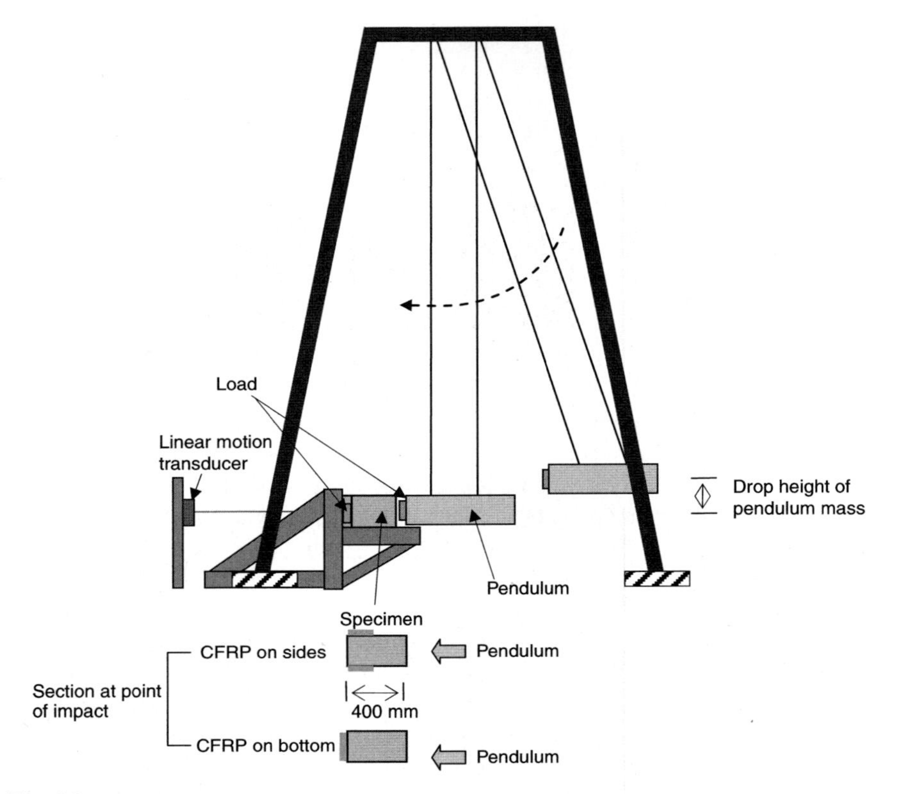

Fig. 6 Impact test setup

Table 2 Test results under impact loading

Test	Layout of CFRP repair	Concrete strength, Mpa	Surface preparation	Failure mode
Bm 1	No anchors	35	Sand blast	Sheet Delam. (Fig. 7)
Bm 2	6 anchors each side Anchor material = Sheet material	14	Sand blast	Sheet Fracture (Fig. 8)
Bm 3	Same as Bm 2	35	Sand blast	Sheet Fracture
Bm 4	Same as Bm 2	14	No adhesion	Anchor Fracture (Fig. 9a)
Bm 5	30% more CFRP than in Bm 2	35	No adhesion	Sheet Fracture (Fig. 9b)
Bm – Col 1	4 anchors each side of column Anchors 610 mm from column face	42	Grinding	Sheet fracture after steel yielding (Fig. 10a)
Bm- Col 2	6 anchors each side of column Anchors 330 mm from column face	42	Grinding	Steel yielding and hinge development (Fig. 10b)

Fig. 7 Delamination failure of Beam 1

Fig. 8 Fracture of CFRP sheet in Beam 2 (See also Color Plate 2 on page 210)

a) Anchor failure, Beam 4 b) Fracture of sheet, Beam 5

Fig. 9 Failure with no adhesion between CFRP sheet and concrete (See also Color Plate 3 on page 210)

a) Sheet fracture, Bm-Col 1

b) Hinging and fracture, Bm-Col 2

Fig. 10 Failures of Beam-Column specimens (See also Color Plate 4 on page 210)

2.2 Cyclic Loading of Columns with Inadequate Splice Lengths

2.2.1 Overview

As indicated earlier, the efficiency of the confinement provided by a rectangular steel jacket can be increased and the behavior of lap splices away from the corners can be improved. Similarly, it was envisioned that CFRP anchors may be used with or without CFRP jackets in strengthening lap splices in rectangular columns with walls. The objective was to develop effective methods of strengthening deficient lap splices in rectangular reinforced concrete columns by a combination of CFRP jackets and CFRP anchors or using CFRP anchors only. A series of tests were conducted under monotonically increasing loads to failure to optimize the anchor details. Three rectangular (460 mm × 910 mm) columns were fabricated and strengthened by CFRP anchors with or without CFRP jackets. Column 1 was tested as-built under monotonic loading and provides a reference against which to compare the behavior of Column 2 and 3 that were tested under cyclic loading after strengthening with CFRP sheets and anchors.

2.2.2 Test specimens

The geometry and loading configuration for the test specimens are provided in Fig. 11. The longitudinal bars in the column and the bars from the footing were lap spliced above the construction joint between the column and the footing. The footing was fixed to a strong floor by threaded rods and lateral load was applied to the column at 2740 mm from top of the footing. No axial force was applied to the columns. To minimize effect of shear, the columns were designed to have higher nominal shear strength than flexural strength. The test columns had twenty lap splices (10 on each face). The length of the #8 (25 mm) lap splices was 610 mm. In the lap spliced region, transverse reinforcement was provided by #3 (10 mm) bars

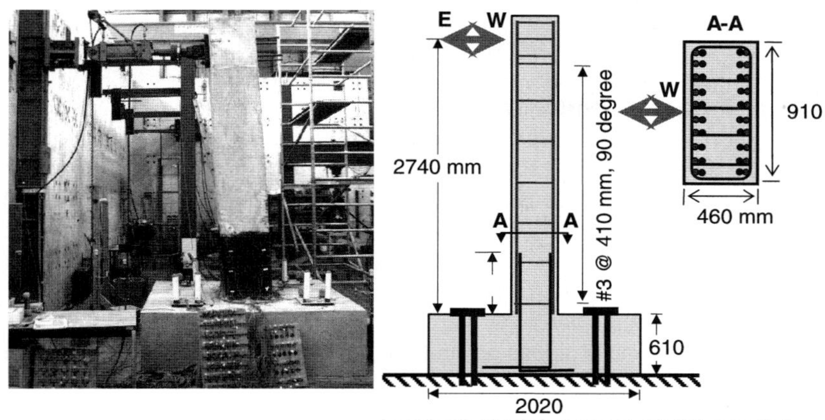

Fig. 11 Test setup and specimen

at 410 mm spacing with the first tie at 100 mm from the footing. Design of columns was based on provisions of the ACI 318–63 (Aboutaha et al., 1999).

Design compressive strength of concrete was 28 MPa. The measured compressive strengths of concrete at the day of the test are shown in Table 3. The steel reinforcement used for the tests was GR60 (414 MPa, tensile yield strength) and the measured tensile yield strength was 434 MPa for all the longitudinal reinforcement. One mm thick CFRP sheet (Tyfo® SCH-41 composites with Tyfo® S Epoxy) was used in column strengthening. The ultimate tensile strength was 1000 MPa at a strain of 0.01.

Table 3 Specimens and strengthening methods

Column	Test Condition	CFRP Jacket		Anchors			
				No. (A)	Hole Dia. mm	Width of CFRP mm (B)	Total Width mm ($A \times B$)
Col 1 ($f_c' = 32$ MPa)	As-built (Monotonic loading)						
Col 2 ($f_c' = 39$ MPa)	Strengthened (Cyclic loading)	West	Full wrap	8	19	180	1440
		East	Full wrap	16	13	90	1440
Col 3 ($f_c' = 39$ MPa)	Strengthened (Cyclic loading)	West	None	20	16	130	2600
		East	Partial wrap	16	13	90	1440

2.2.3 Strengthening Methods

- *Column 1* No CFRP jackets were applied to Column 1.
- *Column 2* The CFRP jacket for Column 2 is shown in Fig. 12. One layer of CFRP sheet was used to provide confinement to the 610 mm lap spliced region (Two 305 mm wide CFRP sheets were used). The CFRP sheet was overlapped by 130 mm on one side face of column after wrapping (Fully wrapped CFRP jacket).
- *Column 3* Column strengthening is particularly difficult if infill walls are present. It was assumed that 305 mm wide walls framed into the north and south faces of the column and were even with the west face of the column. Therefore, CFRP jackets could not be wrapped around the column. Figure 13 shows the jacket details. One layer of CFRP covered the east face and 150 mm of the north and south face up to the wall. The short sides of the jacket were anchored by four CFRP anchors (Partial CFRP jacket). On the west face, one layer of CFRP was applied to the face before applying the anchors to provide a more uniform distribution of confining force from the anchors but this sheet by itself did not provide any confinement to the splices without the anchors.

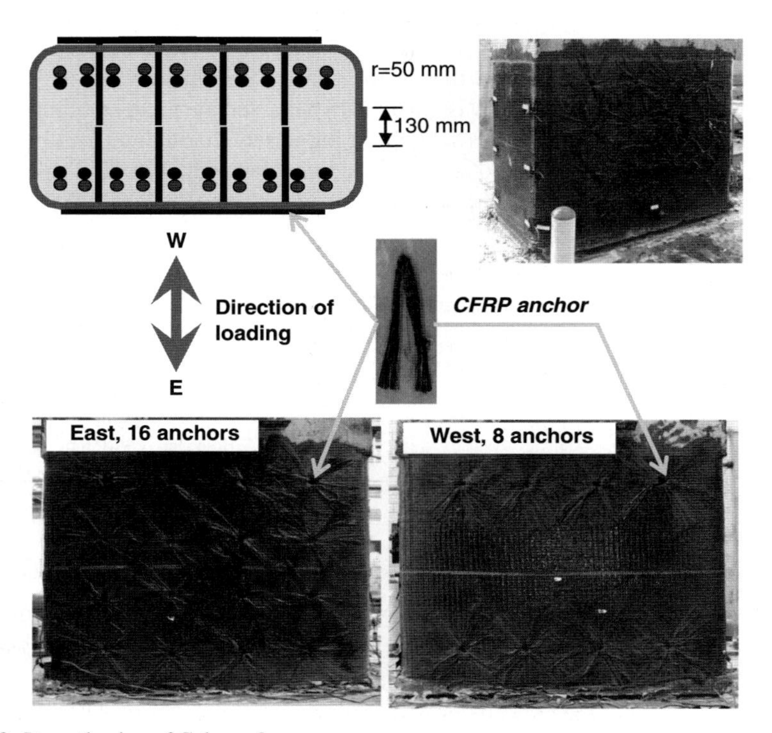

Fig. 12 Strengthening of Column 2

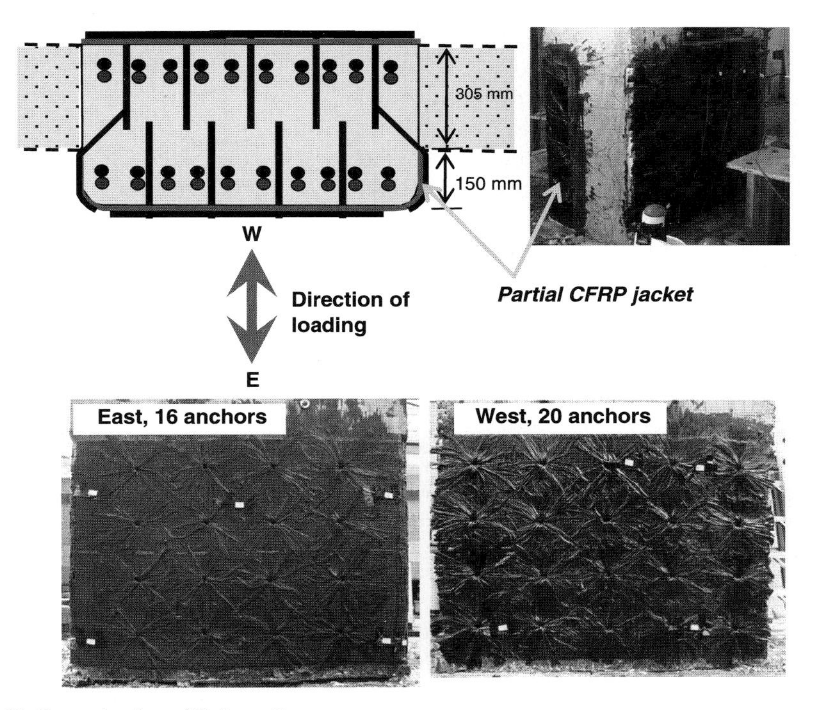

Fig. 13 Strengthening of Column 3

Cement paste was removed from concrete surface of a column where CFRP jackets were applied. The corners of all the test columns were rounded to a 50 mm radius to make a smooth transition of CFRP around a corner. CFRP anchors as shown in Fig. 2 were used. CFRP anchors in the test columns are summarized in Table 3 and the geometry of CFRP anchors in the tested columns is shown in Figs. 12 and 13.

For Column 2, the total width material in the CFRP anchors in both sides of column was the same. However, 16 anchors were applied to the east face while 8 anchors were applied to the west face. The width of CFRP per one anchor in the west side was twice that in the east side. The geometry of CFRP anchors on the east face of Column 3 was identical to that of Column 2. Twenty CFRP anchors were applied to the west face of Column 3 since there was no confinement provided by wrapping a sheet around a corner.

2.2.4 Test Results

Figures 14 and 15 show drift ratio vs normalized load responses.

The lateral load applied to the test columns was normalized using the computed nominal strength (250 kN) of the column. The nominal strength was calculated using the design strength of the concrete and reinforcement. Column 1 was tested as-built under monotonic loading to determine the strength of the test specimen without

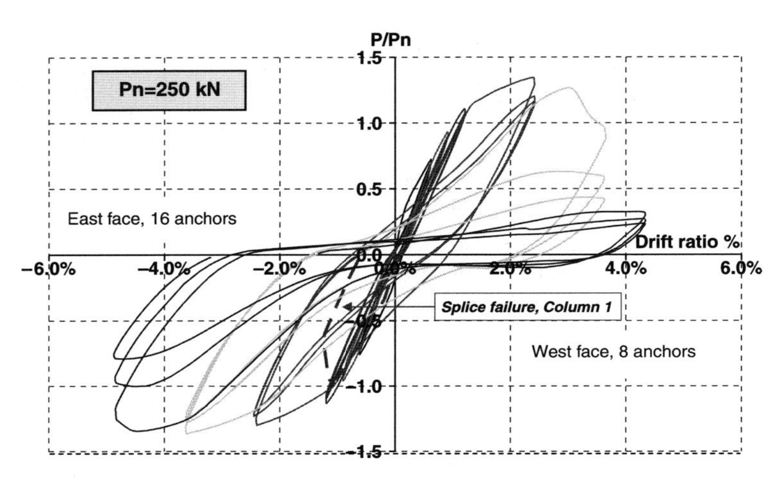

Fig. 14 Drift ratio vs normalized load, Column 2

strengthening and is shown for comparison with the cyclic responses of Column 2 and 3 in Figs. 14 and 15.

A summary of the column test results is shown in Table 2. The failure mode of Column 1 was a brittle splice failure before the nominal strength of the column was reached. The failure mode of the columns after strengthening was yielding of tension reinforcement. Significant improvement of strength and deformation capacity under cyclic loading was observed in both directions for both strengthened columns. Although there was little difference in strength as a function of the number of anchors, the drift ratio corresponding to the peak strength of Column 2 was 3.6% on the east face and 2.4% on the west face. For Column 3, both directions showed

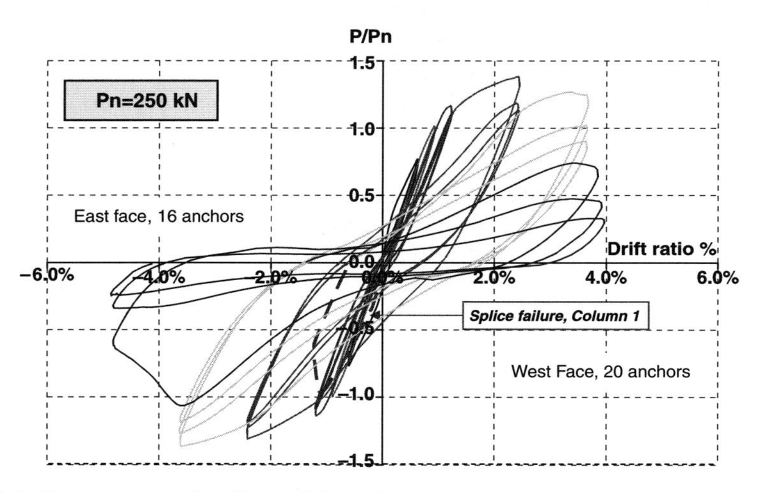

Fig. 15 Drift ratio vs normalized load, Column 3

Table 4 Specimens and summary of test results

Col.	Test condition		CFRP jacket	No. of anchors	Meas. peak strength, P/P_n	Drift at peak strength
1	As-built				0.96*	1.1%
2	Strengthening	West face	Full wrap	8	1.35	2.4%
		East face	Full wrap	16	1.36	3.6%
3	Strengthening	West face	No wrap	20	1.38	2.4%
		East face	Partial wrap	16	1.36	3.6%

*Splice failure

considerable improvement compared with Column 1 even though the column was not fully wrapped.

3 Concluding Remarks

Beams strengthened with CFRP sheets and anchors to provide continuity of positive moment reinforcement performed well under impact loadings. The failure modes were identical to those of specimens subjected to static loading. Through the use of anchors it was possible to fracture the CFRP sheets and fully utilize the strength of the carbon fibers.

A brittle splice failure occurred in the as-built rectangular column which was designed based on provisions of the ACI 318-63. The as-built column exhibited little or no ductility before splice failure occurred. However, the columns strengthened with CFRP showed a significant increase in strength and deformation capacity under cyclic loading compared with the as-built column. Deformation capacity was improved by using a larger number of smaller anchors. The use of a partial CFRP jacket or a CFRP sheet on one face improved the deformation capacity less than when using fully wrapped CFRP jackets. However, the same improvement in strength was achieved using partial jackets.

Acknowledgments This work was supported under a grant from the National Science Foundation. FYFE Co. LLC provided CFRP materials. The authors would like to thank Sarah Orton for her participation in the project and her assistance in the tests reported here. Loring A. Wyllie, Jr., Dilip Choudhuri and Viral B. Patel provided many useful suggestions and comments during the course of this project.

References

Aboutaha R S, Engelhardt M D, Jirsa J O, Kreger M E (1999) Experimental Investigation of Seismic Repair of Lap Splice Failures in Damaged Concrete Columns. ACI Str Jl 96, 2, 297–306

Estrada J I (1990) Use of Steel Elements to Strengthen a RC Building. MS Thes U Austin, TX

Kim I S (2008) Use of CFRP to Provide Continuity in Existing RC Members Subjected to Extreme Loads. PhD Diss U TX Austin

Kim I S (2006) Rehabilitation of Poorly Detailed RC Structures Using CFRP Materials. MS Thes U TX Austin

Kim I S, Jirsa J O, Bayrak O (2008) Use of CFRP to Strengthen Lap Splices of RC Columns. IABSE Cong Chicago

Kobayashi K et al. (2001) Advanced Wrapping System with CF-Anchor. Proc 5th Int Conf on Fibre Reinf Plastics for RC Struct V1: 379–388

Orton S (2007) Development of a CFRP System to Provide Continuity in Existing RC Buildings Vulnerable to Progressive Collapse. PhD Diss U Austin, TX

Orton S, Jirsa J O, Bayrak O (2006) Anchorage of CFRP Sheets with and Without Height Transition. Third Int Conf on FRP Composites in Civ Engr CICE 2006, Miami: 665–668.

Özdemir G, Akyüz U (2006) Tensile Capacities of CFRP Anchors. Advances in Earthquake Engineering for Urban Risk Reduction Springer: 471–488

Earthquake Risk Mitigation – The Global Challenge

Robin Spence

1 Inroduction

Although not yet ended, it is already clear that the present decade is going to be worse than perhaps any in the last century in terms of earthquake casualties. The cumulative death toll from the events of this decade, including Bhuj, 2001, Bam, 2003, South Asia 2004, Pakistan, 2005 and Wenchuan 2008 already exceeds 520,000[1]. And while earthquake death tolls are the most clearly observable measure of the impact of an earthquake, the accompanying physical damage and economic losses in each of those events, all of which occurred in relatively poor regions of the world, has had a devastating impact on the progress of economic development in the affected areas.

In 2000, the UN Millenium Summit, in New York, adopted the so-called Millenium Development Goals (MDGs), *UN Millennium Declaration (A/RES/55/2).* the aim of which is, by 2015 to radically reduce poverty in the world. This will be achieved by a programme targetted on eradicating extreme poverty and hunger, achieving universal primary education, promoting gender equality and ensuring environmental stability; with a view to putting all parts of the world on a path to self-sustaining development (Sachs, 2005a). As initially formulated, these MDGs do not specifically identify natural disaster risk reduction (DRR) as a component of the strategy, but it has become clear, and is acknowledged in many subsequent UN documents, that the vulnerability to natural diasters constitutes one of the major obstacles to the achievement of the Millenium Development Goals. Figures 1 and 2, from the report of the UN Millenium Project (Sachs, 2005b), illustrates diagrammatically how, from 2005, it was feasible to upscale the action plan towards the

R. Spence (✉)
Cambridge Architectural Research Ltd., 25 Cwydir Street #6, Cambridge, CBI 2LG, UK
e-mail: r.spence@carltd.com

[1] Based on the Wikipedia list of fatal earthquakes since 1900, Trevor Allen. This depends for years up to 2000 on the Utsu catalogue, and includes the lower, official figure of 253,000 for the 1976 Tangshan earthquake.

A.T. Tankut (ed.), *Earthquakes and Tsunamis,* Geotechnical, Geological, and Earthquake Engineering 11, DOI 10.1007/978-90-481-2399-5_4,

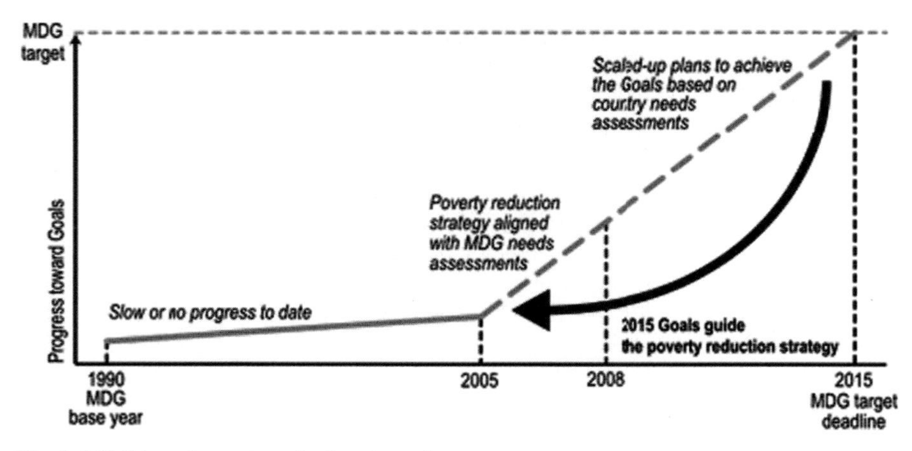

Fig. 1 MDG-based poverty reduction strategies

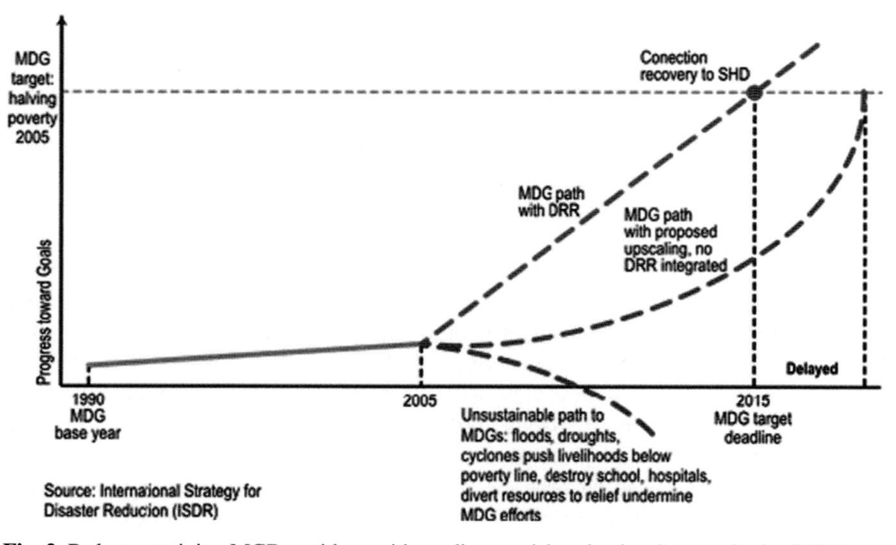

Fig. 2 Paths to attaining MGDs: with or without disaster risk reduction Source: Sachs, 2005b

achievement of the MDGs; but only with DRR built into all development planning, especially where the construction of supporting infrastructure is involved.

To assist this process, the then UN Secretary General, Kofi Annan proposed the establishment of a "worldwide early warning system for natural hazards" building on existing national and regional capacity. Such a worthy aim, conceived in the aftermath of the South Asian tsunami and several devastating tropical cyclones, might be valuable to save lives in such types of disasters, but it would not be effective to protect property from destruction; and in the case of earthquakes, no reliable means of earthquake prediction yet exists. During the last few years the evidence of development set-backs through earthquake losses has become very clear, for instance in Sri Lanka where the reconstruction cost was more than 10% of the country's entire GDP in that year (EEFIT, 2006)., in Pakistan where the World Bank estimated that rehabilitation costs would eventually exceed $3.5 bn (EERI, 2006) about 3% of its GDP in that year, and now in China where it has been reported (EERI, 2008) that almost $150 bn, more than 25% of a single year's tax revenue for the whole of China, will have to be allocated to replace the losses in the May 2008 earthquake, with equivalent reductions in planned development expenditure.

A particularly poignant aspect of many of the recent tragedies is that many of the victims were school children, who died because their schools had insufficient resistance to withstand the earthquake shock. Consider the following recent examples, reported by the Disaster Education Network:

- The Kashmir earthquake in Pakistan of October 2005 killed at least 17,000 students in schools and seriously injured another 50,000, leaving many disabled; over 300,000 children were affected. Moreover 10,000 school buildings were destroyed; in some districts 80% of schools were destroyed.
- In the Wenchuan earthquake in China in May 2008, more than 7,000 children were killed in their schools and an estimated 7,000 classrooms were destroyed
- In the $M = 6.4$ Baluchistan province earthquake in October 2008 school buildings were among the structures worst affected; according to local reports the earthquake destroyed 98% of educational institutions in Ziarat, 50% in Loralai and Pishin, and 25% in Quetta. There were about 150 primary, secondary and high schools in Ziarat.

Even if we do not have precise numbers, it seems that schoolchildren have been disproportionately represented amongst the victims in the Boumerdes earthquake (Algeria, 2003), the Kashmir earthquake (Pakistan, 2005) and the Wenchuan earthquake (China, 2008) as well as in the 2003 Bingol earthquake in Turkey. There could be no clearer example of an obstruction to development than the avoidable death of a child as s/he is learning the skills to contribute to that economic development for the community.

But a clear and achievable global strategy to reduce and eliminate such devastating losses does not appear to be in sight. These events have come at the end of a half century during which substantial progress in earthquake science and technology have been achieved: the reasons for earthquake risk have become very well-understood, mapping of the earthquake risk has become increasingly more refined, codes of practice for building in earthquake areas have become more detailed and comprehensive, and techniques for mapping and estimation of future losses have developed rapidly. The International Decade for Natural Disaster Reduction (1990–2000) set out, from 1990, to "reduce catastrophic life loss, property damage and social and economic disruption from natural hazards" (Housner, 1989). What was achieved in the 1990s should not be minimised. Much effort was devoted to improving the mapping of natural hazards, identifying areas of especially high risk, researching low-cost methods to reduce earthquake risk, improving national codes of practice for earthquake-resistant construction, and preparing simple manuals to explain better building techniques to home owners and small builders.

But whatever local success such projects have had, it is now clear that they have made little or no impact on the apparently increasing surge of earthquake losses. Understanding of risk did not reach those who face that risk; and improved codes of practice (e.g. in Turkey, Algeria, and China) did not eliminate construction of buildings of extremely high vulnerability; and earthquake-resistant design manuals did not affect more than a tiny minority of those who might have benefited from them.

Clearly a new approach is needed. This paper aims to take a fresh look at global earthquake risks, to consider what has been achieved in particular countries, and what are the determinants of success in earthquake risk mitigation. It aims to propose some ways in which the engineering community, working with other professionals, public administrators and politicians, can begin to act in ways that will over time, make a difference.

2 Global Variations in Risk

To make progress in reducing earthquake risk, clearly those area with greatest risk must be identified and targeted. Efforts to reduce risk must also acknowledge that earthquake risk is, by comparison with other causes of death, very low at a global scale. The Global Burden of Disease Survey of 1990 (Murray and Lopez, 1997) showed that, globally, unintentional injuries constituted less than 6% of all deaths, of which about 30% were accounted for by traffic accidents, but only 1.3% were due to all natural disasters, and only 0.3% were from earthquakes. Thus earthquake deaths constituted only 180 in every million deaths on a worldwide basis, an apparently small risk for public health planners, and one which has probably changed little since that time. However, earthquake risk is not uniform, and in some areas of the world it is of course very much higher than this.

It is clear, for example that earthquake risk is concentrated in the poorer countries, and it is instructive to examine comparative annual earthquake death rates by regions of the world according to their income levels (Spence, 2007). Figures 3 and 4 show annual death rates per million population since 1900.

Clearly death rates in the affluent countries have been declining since 1900 and are now very low at 1 per 100 million, while those in the poorer countries have not significantly declined and remain very high at around 15 per million population, a ratio of 1500 to 1. Of course poverty itself, and the accompanying lack of resources and lack of earthquake awareness is a big contributory factor. Chen et al. (2002) has shown that there is a close relationship between earthquake vulnerability and per capita income across many countries in the world, and even within a large country such as China (Fig. 5).

But income disparity does not explain all of the variation in earthquake risk. Other local factors play an important role. One of these factors is where, within an earthquake region, people choose to build. In Iran (which, with an annual death rate exceeding 2 per thousand over the last 25 years has the highest rate of earthquake deaths of any major country), the traditional pattern of settlement can be

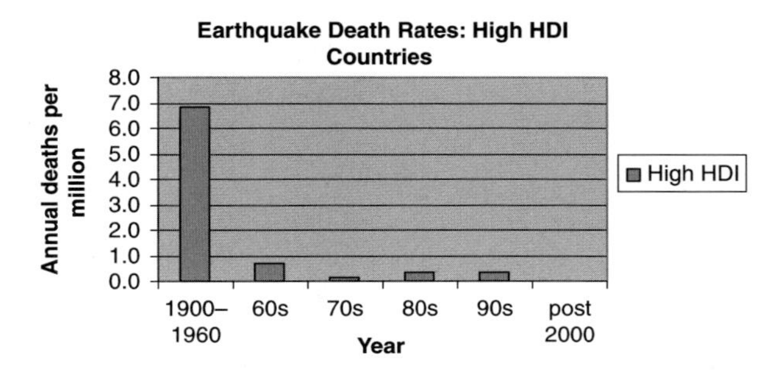

Fig. 3 Annual death rates per million population since 1900: affluent countries. Huge and sustained improvement since 1960 (HDI = human development index)

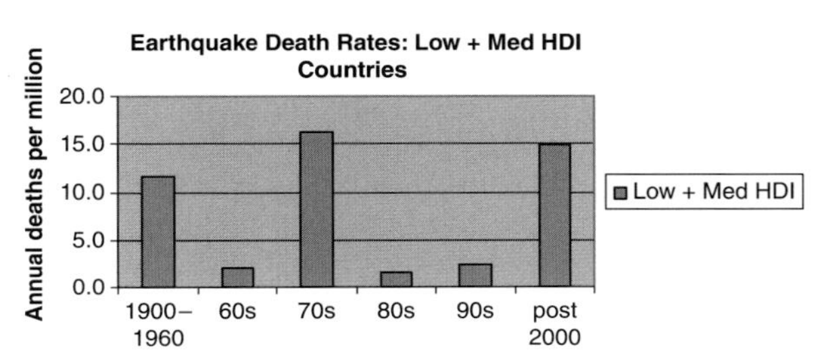

Fig. 4 Annual death rates per million population since 1900: poorer countries. No sustained progress

shown to relate very closely to the location of major earthquake faults (Jackson, 2006). This is no accident. Jackson has shown that, in this dry and mountainous country, the availability of water determines the locations where people can build settlements. Active faults create elevated water tables which have been skillfully

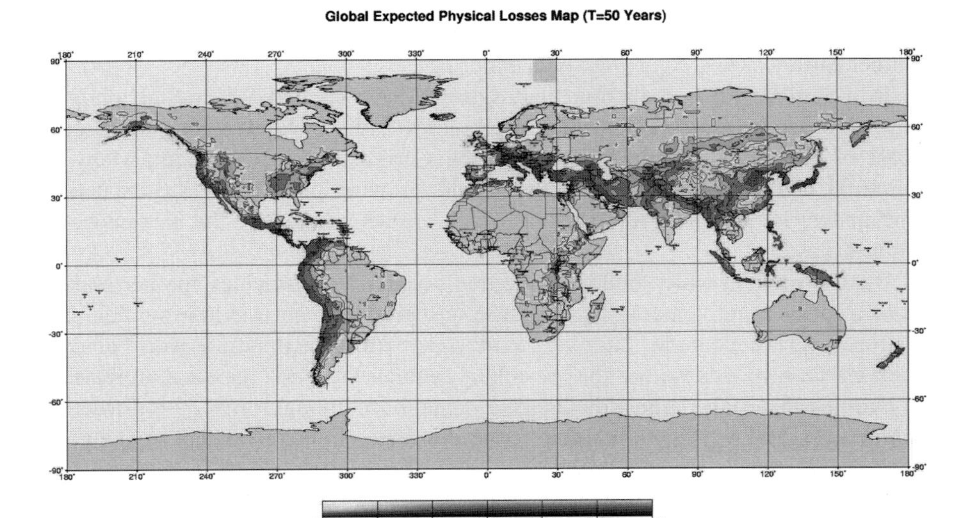

Fig. 5 World map of seismic risk. Expected loss by 0.5 × 0.5 degree, $US bn over next 50 years, derived from income-related vulnerability data (Chen 2002)

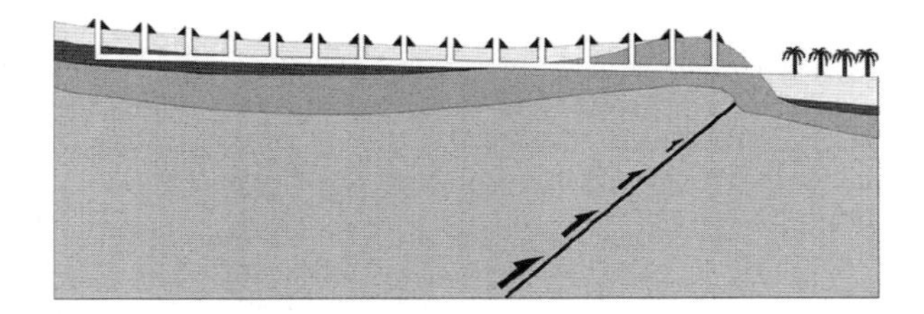

Fig. 6 The development of elevated water table through thrust faulting. Source: Jackson, 2006

exploited through traditional engineering over many centuries to provide irrigation to the desert region, and enable cultivation to take place (Fig. 6).

These irrigated zones have in turn led to the development of large towns, and Jackson (2006) shows how Tabas and Bam, cities devastated in earthquakes in 1978 and 2003, killing high proportions of their populations, have been located close to just such active faults for reasons of irrigation. Tehran, with a population today exceeding 10 million, is also located astride such a fault. As Jackson puts it:

> In such places, earthquakes that in the past killed a few hundred or thousand people will now kill tens or hundreds of thousands or more. The situation is similar throughout much of the Mediterranean-Middle East-Central Asia earthquake belt.

In areas such as these where the mountainous topography is largely created by earthquake fault movement, settlements tend to concentrate on the mountain edges because these locations are on the trade routes, or near to water supplies; and these are the very locations most likely to be hit directly by large earthquakes. Both the 2005 Kashmir earthquake in Pakistan and the 2008 Wenchuan earthquake in China, the two most devastating earthquakes of recent years, struck such regions. Yet, although the deaths tolls were huge, they could easily have been much worse; neither earthquake directly affected a major city.

In a similar way, subduction fault systems have created the coastal topography which encourages settlement and promotes trade, in Japan, along the Pacific coast of North and South America, in the Caribbean and elsewhere. Many of the areas of the world with the most rapidly-growing populations are located in earthquake-risk regions – for reasons which are determined by the prevalence of earthquakes.

But the ways in which buildings are traditionally built provides another pointer to areas of extreme vulnerability. Because of their dominant role in providing protection from the elements, buildings tend to be built throughout the world using a form which responds well to the prevailing climate. This, and the local availability of building materials, largely dictates the range in traditional forms of construction found across the world; but building forms developed in this way may have a good or a poor resistance to earthquakes. Lee, 2008) has undertaken a study of the relative climatic and seismic responsiveness of the range of global building forms, and has matched a preliminary global mapping of traditional building types, based primarily on a recent USGS study (Jaiswal and Wald, 2008), in relation to their seismic and their climatic appropriateness. The comparative mapping is shown in Figs. 7 and 8.

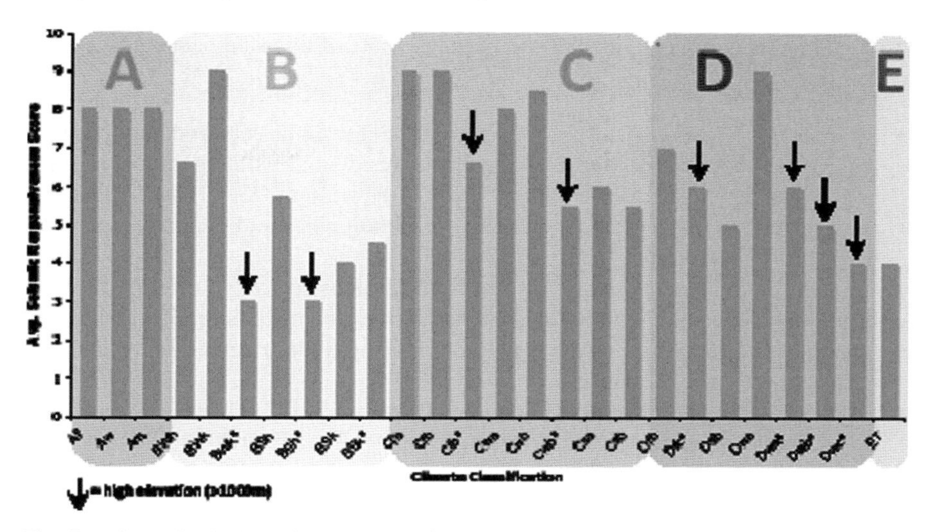

Fig. 7 Estimated seismic resistance index for buildings designed with appropriate response to each of the worlds major climate zones (A=equatorial, B= arid, C= warm temperate, D=snow, E=polar; seismic index 0–3 = poor, 3–7 = moderate 8–10=good) Source Lee, 2008. Note general poor expected performance of buildings suitable for arid climates

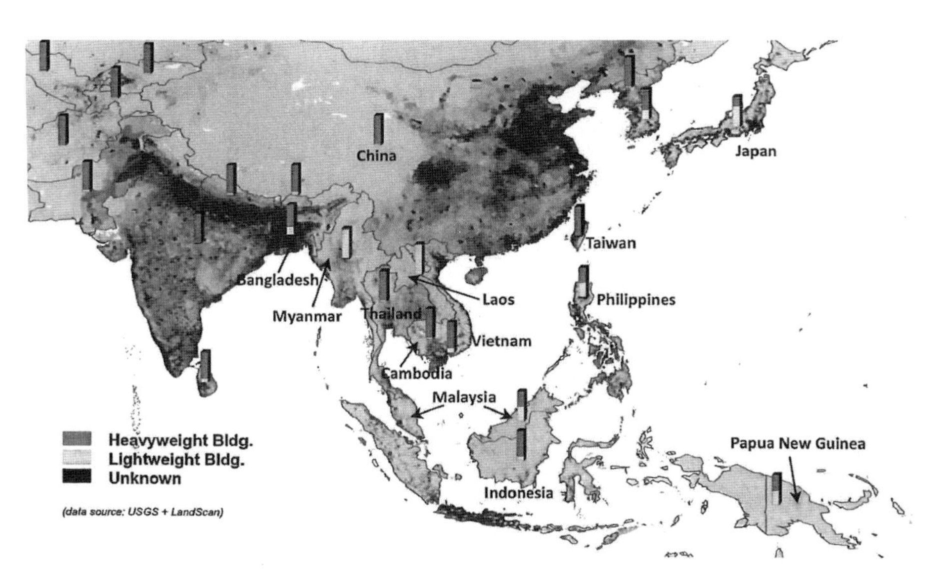

Fig. 8 Distribution of main building types (USGS) mapped onto population distribution (CEISIN data) for South and East Asian counries (Source: Lee 2008)

This kind of work is in its infancy, but the results of this study are significant. It shows that

- In a few of the seismically active regions, lightweight construction – usually based on timber frames – predominates for residential buildings: most of these regions are among the richer nations – Japan, the western USA, New Zealand, but also include poorer countries including the Philippines and Papua New Guinea and parts of Indonesia.
- In most of the world's most seismically hazardous regions – the Middle East, Central Asia, Central and South America, heavyweight construction dominates, leading to a greater risk of the construction being of high vulnerability, and also lethal when it does collapse.
- Thus in most parts of the world dominant forms are likely to have poor resistance to earthquakes: seismic hazards do not strongly affect the way people build

In most earthquakes which create large proportional losses and high death tolls there is a combination of explanatory factors: active faults which have encouraged settlement; a climate which encourages the use of heavyweight materials; and low income levels (often less than $3 per day), which mean that people do not have the means to build in more resistant form. Urbanisation may also, in certain circumstances, make matters worse, but it can be an opportunity for building better buildings in the relatively near future. By developing a better understanding of the regions in which these factors combine, it is possible, with tools we already have available, to predict where disasters are likely to occur in the near future, even if we cannot predict the timing of large events.

3 Relative Progress

Some countries have been more successful than others in reducing earthquake risks, and it is instructive to examine these differences and also to identify reasons for them. In 2006/2007 a survey was undertaken by the author of a small selection of experts from different earthquake-risk countries to obtain an assessment of each country's achievements in earthquake risk reduction over the last 50 years. The questions asked were

1. In your view what have been the most significant successes and failures of earthquake protection in your country during the past 50 years?
2. How successful has been the implementation of new codes of practice in the design and construction of new buildings ? What have been the major obstacles?
3. What do you estimate is the proportion of unsafe buildings (ie built before current codes, or without applying codes) in the current building stock?
4. Are there any programmes in place or in development to assess and upgrade unsafe buildings (public buildings – e.g. schools and hospitals; residential buildings) and how successful have these programmes been?

Responses were obtained from 31 experts in 22 different counties, and these have been reported in detail elsewhere (Spence, 2007). On the basis of the responses it seemed possible to divide the counries into 4 separate groups.

In the first group are what we could describe as the "success stories". These countries have made demonstrable progress in tackling their earthquake risk and reducing it to a level much lower than existed 50 years ago; in each country there remain problems to be tackled, but there has been measurable progress. Japan, California and Eastern North America, and New Zealand belong to this group. A second group of countries has made some progress over the last 50 years: but it is relatively slow and limited progress. Overall risks are lower than they were, but many high risks remain, and much remains to be done to raise the awareness of the public and the government in order to tackle earthquake risk in a sustained and effective way. In this category of "slow progressers" are most of the European countries. A third group of countries may be described as "movers"; starting from a relatively high level of risk until recently, much has and is being done to raise awareness, and to tackle the high risks which exist in the country: China, Colombia, Turkey and several Caribbean countries belong to this group. A final, large and most worrying group are those countries in which risks are already high, but, in spite of the best efforts of a few dedicated professionals, little is being done at a national level to tackle the legacy of risk or to control its causes, and consequently risks are continuing to rise alarmingly. In this "growing risk" category are Iran, India, Nepal and Algeria and several other countries.

In Figs. 9, 10 and 11, the responses to some of the questions have been quantified and are summarised in graphical form by country and country group. Figure 9 shows relative levels of code implementation, as reported by the responders. There is clearly a wide variation, with the "success stories" reporting mainly good implementation levels, exceeding 70%, while at the opposite end the "growing risk" category reported uniformly poor code implementation. Figure 10 reports the respondents' estimates of the proportion of unsafe buildings in the country's building stock, in some ways the inverse of Fig. 9, but including of course all the existing building stock. All countries reported some unsafe buildings (country understanding of what constitutes an unsafe buildings also of course varies). But it is worth noting that even in the "success stories", up to 10% of the country's buildings are regarded as unsafe, while among the growing risk category, it was up to 90%. Finally, Figure 11 reports the experts views on the progress of identification and strengthening of unsafe buildings. A scale was developed with 4 categories:

Category 1: extensive retrofits carried out
Category 2: retrofits to some schools, hospitals
Category 3: a few demonstration projects
Category 4: a few assessments only

On this criterion, none of the countries, even the success stories, were found to be in the highest category, but again, there were of course marked disparities between the four country groups.

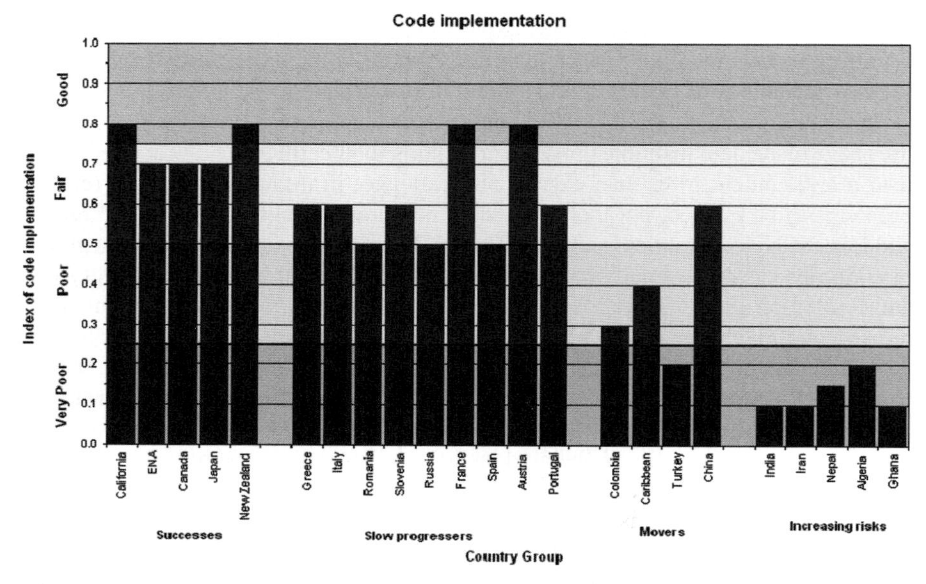

Fig. 9 Code implementation: a country and country group comparison

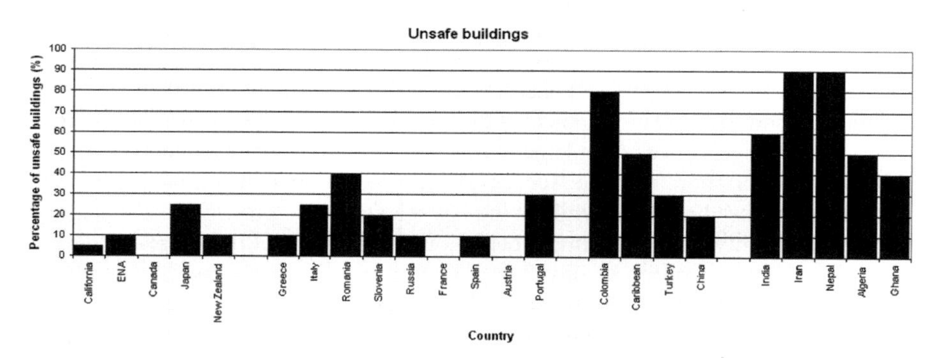

Fig. 10 Proportions of unsafe buildings: a country and country group comparison

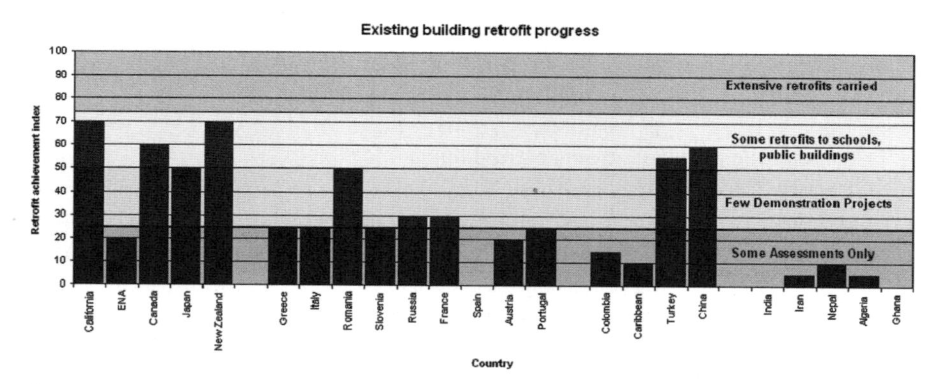

Fig. 11 Progress in retrofitting existing buildings: a country and country group comparison

These responses, and the information reported with them, also enabled some general conclusions about the reasons for the differences to be made. There are two strong determinants of action for earthquake risk mitigation: the first is recent experience of a strong or devastating earthquake, and the second is the availability of resources to take action for mitigation. All of the "success stories" are relatively affluent countries (in the top of 10 in the UN's Human Development Index, HDI), while the "growing-risk countries" are towards the bottom of the HDI list. Europe, while its level of affluence is not far short of that of the "success stories", is doing relatively little because large earthquakes have been few, and public perception is that the risk is small and localised. The "movers" are countries with growing economies that provide opportunity for national action. In spite of these economic and tectonic determinants there are some lessons which the slow-progressers and high-risk countries can maybe learn from the "success-stories" and the "movers".

1. That public awareness, rather than either building codes or law, is everywhere the most important basis for action.
2. That education, training and registration of professionals are vital ingredients of success.
3. That the experience of damaging earthquakes has given great impetus to progress; it seems vital to be prepared with changes in regulations and practices which can then be rapidly implemented in this moment of opportunity.

Two common concerns, shared by virtually all the lower-income countries, are:

- That their uncontrolled urbanisation means that the vast majority of the new building stock is built without any concern for earthquake risk. Attempts to create a system of building control are hampered by corruption, and by the attitude that the issue of building permits is a revenue-generating process, not connected with public safety, and this is building up the potential for huge disasters in the future.
- Where the need for strengthening of key public buildings such as schools and hospitals is identified, and shown to be feasible, resources to undertake it are not made available.

4 New Initiatives and Prospects

To make progress in earthquake risk reduction, some kind of comprehensive global strategy is needed. There have been efforts, spearheaded by Geohazards International and IAEE to set up some kind of international organisation with this remit (Tucker, 2004). The European Association for Earthquake Engineering has submitted to the European Union a proposal for a programme for earthquake risk mitigation in Europe; and other country efforts have been made (Spence et al., 2005). In the US the National Earthquake Hazards Reduction Programme, in place since 1977, has had some success (EERI, 2003).

It is valuable to learn from public health programmes, which over timespans of decades have created radical improvements in the risk from disease. The US Centre for Disease Control has identified three types of action needed for successful programmes of disease control, which can be characterised as:

1. Wide application of known technologies
2. Development and application of new technologies
3. The creation of a safety culture.

Each of these types of action has relevance in approaching the problem of global earthquake risk reduction – which can indeed be considered as a type of public health programme. Some brief examples will be given.

4.1 Application of Known Technologies

A great deal is now known about how to design and build buildings safely in earthquake areas, and how to utilise reinforced concrete, masonry, steel, timber and other materials for this purpose. Virtually every country has its earthquake engineering specialists, trains engineers in how to use the common engineering materials, and has a Code of Practice for construction in earthquake areas. One part of the solution is to find better ways to constrain homebuilders to build according to the codes. In Turkey and in India, as elsewhere, following recent devastating earthquakes, there has been much good work to find ways to improve the standards of building control (Gulkan et al., 1999; Jain, 2006). In Turkey, the national earthquake insurance scheme has given an impetus to this effort, and similar schemes have been or will be will be adopted in other countries (e.g. Romania, Algeria) in the near future.

To achieve better standards of building and Code compliance, it is essential that proper systems of licensing engineers are put in place so that those who have responsibility for all buildings are competent; and that engineers are held to be liable for the performance of their buildings in earthquakes, which means, among other things, that they will have to take some responsibility for site supervision. This is partly a matter for government, but professional engineering bodies in all earthquake countries need to be involved in defining the appropriate laws and campaigning to have them passed into statute.

Neverthess, these Codes are applicable essentially to new buildings in modern materials, buildings which may be few outside the cities and large towns, where peoples houses are built as a response to the climate and using whatever materials are available. Building for safety programmes have been sucessfully carried out in many rural areas which have been devastated by earthquakes, in Yemen, Ecuador, Peru, the Latur district of India, and in rural Pakistan. These have been supported by training programmes for rural builders, and excellent training manuals have been produced. Sadly, few of these programmes have to date, been extended into the areas adjacent to those affected by the most recent earthquake, which may be even more likely to be affected by the next one.

4.2 Harnessing New Technology

The great advances in the technology of earthquake engineering over the last half century have unfortunately made little impact on the way in which the dwellings and workplaces of most of the world's population are built. Active and passive control systems, base isolation, new materials and new techniques for design are neither designed for, or affordable by, those who currently are most vulnerable to earthquakes. Little work to date has addressed the problem of finding and demonstrating low-cost techniques suitable for upgrading low-strength masonry or poor-quality reinforced concrete. Nevertheless work in Ankara, Peru, Wellington, New Zealand and Tokyo has begun to address these problems, with significant achievements demonstrated in the laboratory (Blondet et al., 2006, Mayorca et al., 2006; Charleson, 2006; Schilderman, 2004).

With all such studies, the problem is how to achieve application by the intended beneficiaries, the urban and rural house-builders. This is a problem even when technical innovations are intended for use by highly educated designers working in a modern industry. It is far greater when they are intended to be adopted by individuals in the building of their own houses, who are not easily reached by any formal rules or guidance documents and are likely to be reluctant to pay for strengthening against a threat that may or may not materialise within the next few decades. Extensive builder training and educational programmes will be needed, of the kind which today happen only during the reconstruction programme after a major earthquake.

Loss estimation is an essential tool to identify areas at risk, plan and justify mitigation expenditure, as well as to support insurance schemes, providing the kind of information on likely losses which can stimulate individual and collective action. Loss estimation methods have been developing rapidly in recent years, although many uncertainties remain, especially in our ability to predict ground motions from a given event, estimate the resulting damage, and produce reliable estimates of casualties (Spence et al., 2008). Much research remains to be done, not least to obtain a better understanding of rapidly expanding national inventories of buildings and other facilities at risk. New remote sensing tools have potentially an important part to play in this effort, Fig 12 (Saito and Spence, 2008). The estimation of likely casualties from earthquakes is also poorly understood, but new studies are improving our understanding in this area also (So et al., 2008). However, much of the progress in loss modelling to date has been done by commercial modelling companies, whose models are often seen as "black boxes", producing loss data without explaining the methods (Bommer et al., 2006). New, open-source software models are being planned which could benefit administrations and insurers needing results, but unable to afford the commercial models. In 2008, a new programme for a Global Earthquake Model (GEM) was launched, with the aim of building an independent standard for modelling, monitoring and communicating earthquake risk worldwide. Potentially it will be a powerful tool to raise risk awareness, contribute good information for risk mitigation planning, and help post disaster development.

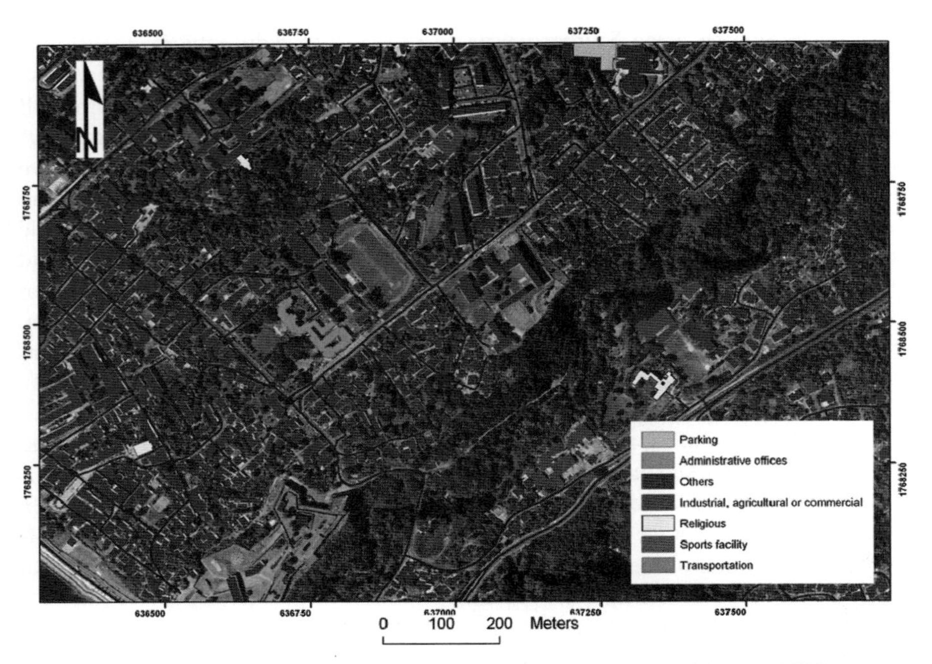

Fig. 12 Use of remote sensing to identify buildings by use type (Saito and Spence, 2008)

4.3 Creating a Safety Culture

None of the mitigation actions which the engineering and scientific community identifies as being needed are likely to succeed without prior efforts to promote, at all levels of society, an awareness of the potential threat from earthquakes, and the belief that something can be done about it. In most parts of the world, damaging earthquakes occur only infrequently, with recurrence intervals of half a century or more; and in this situation public awareness of the earthquake risk is likely to be very low. This makes it essential for the scientific and engineering community, who have a better understanding of the hazard and of the possible impact of a major, foreseeable event, to become involved in activities to raise public awareness. This can involve education and outreach at all levels, from schoolchildren, through builders and property-owners to business and political leaders, and many kinds of activities can be used. Some good models of such programmes have been developed over recent years, in Italy, Nepal and India, (Dixit et al., 2000; NICEE, 2006; Brasini et al., 2006).

On any society's safety agenda, the safety of schoolchildren must have a high priority. Yet, as shown in Section 1, public school buildings have collapsed very regularly in earthquakes, even in cases where most other buildings have survived, and many schoolchildren have been killed. And many more schoolchildren have been spared only because their school collapsed outside school hours. There is clearly a need for engineering intervention, beginning with assessment of all schools in earthquake-prone areas, identifying those at risk, and then either demolishing and replacing or strengthening those found to be inadequate, and much good recent research on these topics has been done (Grant et al., 2006; Penelis, 2001). But in order to create the conditions for such programmes to take place, efforts are needed to highlight and quantify the problem, bring it to the top of the action agenda of the responsible agency, and of politicians, and ensure that the funds are made available. The OECD has recently taken a big step in this direction recently with a Recommendation Concerning Guidelines on Earthquake Safety in Schools, adopted in July 2005, (Yelland and Tucker, 2004), which is in the early stages of implementation in about 15 OECD countries.

In the Canadian Province of British Colombia, Families for School Seismic Safety (FSSS) was set up through the initiative of a Vancouver family doctor, who found that her children's school was amongst many public schools at risk of collapse in a foreseeable earthquake (Monk, 2006). FSSS allied itself with the local engineering community, and used lobbyists to bring the issue into the public arena at the time of Provincial elections, asking candidates to pledge, if re-elected, support for a programme of action to reduce the risk. The highly successful campaign used a public health argument, and successfully argued that funds for this public health programme should not have to compete with other projects for a slice of the education budget. In November 2004, the State Premier made a $1.5 billion commitment to getting all schools upgraded in 15 years.

A soundly based earthquake insurance scheme is a key component of the creation of a safety culture, because it encourages the owners of property to understand the risks they face. It is a well-established practice in many types of insurance that the premium paid is lower if the insured can show that various protection measures have been taken – locks and alarm systems for theft insurance, and smoke alarms for fire insurance, for example – and the same approach could in principle be applied to earthquake insurance. Insurers can also demand that minimum standards of design and construction supervision are required before insurance can be offered. Insurance for earthquakes is still available for only a small proportion of householders worldwide, and is inhibited by various national post-disaster compensation schemes. But it is gradually extending, and its growth is greatly assisted by the development of loss estimation techniques.

5 Conclusion

Some conclusions from this brief review of global earthquake risk mitigaion are:

- Earthquake risk is growing, not shrinking, and that growth is concentrated in the uncontrolled new settlements, the cities, towns and villages of the developing countries. It is made up of millions of small- and medium-sized houses, apartment blocks and commercial buildings which have been built without an awareness of the earthquake threat or of how to counter it.
- There is presently no global strategy for resolving this problem, and no international body with responsibility for developing one.

- There needs to be a shift of emphasis in development aid towards mitigation as opposed to relief; and as Wisner has pointed out (Wisner, 2004), efforts at disaster mitigation are frequently frustrated or negated by conflict or political instability.
- The areas and communities at particularly high risk of future disaster can increasingly be identified and could be targetted for action.
- The current pace of reconstruction in many cities in parts of the developing world is an opportunity to get things right, but this will take many decades of committed action at many levels; and regeneration of public buildings such as schools will be much slower
- The effort must start with large public awareness campaigns, with a major extension of the education of building professionals, and with efforts to achieve better building control.
- The engineering and scientific community must become engaged in efforts convince politicians of the need for action, and ally themselves with groups of individuals who are prepared to campaign for change. This will involve engineers and scientists in some unconventional activities and some unusual alliances.
- Research is needed to find simpler and cheaper ways to achieve earthquake resistance, and international collaboration will be necessary to transfer expertise and resources to the communities in the cities which most need it.

If death from infectious diseases can be dramatically reduced by concerted public health campaigns, so too can earthquake losses and casualties; both are entirely avoidable with the technical means at our disposal. It now needs the political will to devote the resources needed to make it a reality

Acknowledgments The author is greatly indebted to the many colleagues who went to considerable trouble to provide responses to the questionnaire in Section 3. In the United States the respondents were Mary Comerio, Fred Krimgold, Tom O'Rourke and Tom Tobin; in Japan, Yutaka Ohta and Charles Scawthorn; in New Zealand Andrew Charleson, David Dowrick and David Hopkins; and in Canada Carlos Ventura. In Europe the respondents were Panayotis Carydis and Antonios Pomonis (Greece), Agostino Goretti, Mauro Dolce and Giulio Zuccaro (Italy), Alain Pecker (France), Alex Barbat (Spain), Rainer Flesch (Austria), Carlos Oliveira (Portugal), Horea Sandi (Romania), Peter Fajfar (Slovenia) Polat Gülkan (Turkey) and Jacob Eisenberg (Russia). From other countries responses came from Aizu Ren and Weimin Dong (China), Tony Gibbs (Caribbean), Omar Cardona (Colombia), Sudhir Jain (India), Amod Dixit (Nepal), Mahmood Hosseini (Iran), Mahmoud Bensaibi (Algeria) and Titus Kuuyuor (Ghana). My apologies are due to each of these generous respondents for the inevitable compression, and for any unintended misrepresentation, involved in summarising and presenting their perceptions of the situation. Thanks are also due to my colleague Emily So for helpful comments on the draft text.

References

ASAG 1996. *ASAG's Intervention Through Technology Upgrading for Seismic Safety*, Ahmedabad Study ad Action Group, Ahmedabad, India

Blondet, M, Torrealva, D, Vargas, J, Velasquez, J and Tarque, N 2006. "Seismic reinforcement of adobe houses using external polymer mesh", Paper 632, *First European Conference on Earthquake Engineering and Seismology*, Geneva, Sept

Bommer, J, Pinho, R and Spence, R 2006. "Earthquake loss estimation models: time to open the black boxes ?", Paper 834 *First European Conference on Earthquake Engineering and Seismology*, Geneva, Sept.

Brasini, F, Modonesi, D, Sidoti, B and Camassi, R, 2006. "Tutti giu per terra – an active course to discover earthquake topics" *First European Conference on Earthquake Engineering and Seismology*, Geneva, Sept.

Centre for Disease Control 1999. "Achievements in public health 1900–1999: control of infectious diseases", *MMWR Weekly*, 48(29), 621–629.

Charleson, A W 2006. "Low-cost improvement to the seismic safety of adobe construction: strips cut from used car tyres", (unpublished paper), School of Architecture, Wellington University

Chen, Y, Chen, Q, Liu, J, Chen, L and Li, J 2002. Seismic Hazard and Risk Analysis, a Simplified Approach, Science Press, Beijing.

Coburn, A W and Spence R J S 2002. *Earthquake Protection*, John Wiley and Sons.

Dixit, A, Dwelley-Samant, L, Nakarmi, M, Pradhanang, S and Tucker, B 2000. "The kathmandu valley earthquake risk management project: an evaluation", *Proceedings, 12th World Conference on Earthquake Engineering*, Paper 1788.

Earthquake Reconstruction and Rehabilitation Authority, 2006. *Guidelines for Earthquake Resistant Construction of Non-Engineered Rural and Suburban Masonry Houses in Cement Sand Moratr in Eartquake-Affected Areas*, ERRA, Islamabad

EEFIT 2006. The Indian Ocean Tsunami of 26 December, 2004: Mission Findings in Sri Lanka and Thailand, Institution of Structural Engineers, London.

EERI 2003. Securing Society Against Catastrophic Earthquake Losses, Earthquake Engineering Research Institute, California.

EERI 2006. The Kashmir Earthquake of October 8th, 2005: Impacts in Pakistan, EERI Special Earthquake Report, EERI Newsletter, 40, 2, 1–8, EERI, Oakland, California.

EERI 2008. The Wenchuan, Sichuan Province China Earthquake of May 12th, 2008, EERI Special Earthquake Report, EERI Newsletter, 42, 10, 1–12, EERI, Oakland, California.

Grant, D, Bommer, J, Pinho, R, Calvi, G, Goretti, A and Meroni, F 2007. "A prioritisation scheme for seismic intervention in school buildings in Italy", *Earthquake Spectra*, 23, 2, 291–314.

Gujarat State Disaster Management Agency, 2002. *Coming Together*. 3rd Edition, GDSMA, abhiyan.gsdma.undp.

Gülkan P, et al. 1999. *Revision of the Turkish Development Law No 3194 and its Attendant Regulations with the Objective of Establishing a New Building Construction Supervision System Inclusive of Incorporating Technical Disaster Resistance-Enhancing Measures* (3 Vols), Turkish Ministry of Public Works and Settlement.

Gülkan, P 2005. "An analysis of risk mitigation considerations in regional reconstruction in Turkey: the missing link", *Mitigation and Adaptation Strategies for Global Change*, 10, 525–540.

Housner, G W 1989. *Coping with Natural Disasters, The Second Mallet-Milne Lecture*, SECED, Institution of Civil Engineers, London.

Jackson, J 2006. "Fatal attraction: living with earthquakes, the growth of villages into megacities and earthquake vulnerability in the modern world Philosophical", *Transactions of the Royal Society*, A, 364, 1911–1925

Jain, S K 2006. The Indian Earthquake Problem, *Current Science* 89(9), 1464–1466.

Jaiswal, K and Wald D 2008. Creating a Global Residential Building Inventory for Earthquake Loss Assessment and Risk Management, USGeological Survey Open-File Report 2008–1160.

Lee, V 2008. Mismatches in climatically and seismically responsive dominant built forms. MPhil Dissertation, Cambridge University Department of Architecture.

Leslie, J and Coburn, A W 1985. *The Dhamar Building Education Project: Assessment*, OXFAM, UK

Mayorca, P, Navaratnaraj, S and Meguro, K 2006. *Report on the State of the Art in the Seismic Retrofitting of Unreinforced Masonry Houses by PP Band Meshes*, International Centre for Urban Safety Engineering, Institute of Industrial Science, University of Tokyo, Japan.

Monk, T 2006. "Community Involvement on the Road to School Seismic Safety in British Colombia", *Conference on Development in Earthquake Mitigation since the Kobe Earthquake*, Japan Society for Earthquake Engineering, Kobe Japan, Jan 2005

Murray C J and Lopez A D 1997. "Mortality by cause for eight regions of the world: Global Burden of Disease Study", *The Lancet*, 349, 3.5.1997, 1269–1276.

NICEE, 2006. *Newsletter*. National Information Centre of Earthquake Engineering, Kanpur India (www.nicee.org)

OECD 2004. Expert Group Report on Earthquake Safety in Schools Part V in Yelland, R and Tucker, B (eds.) *Keeping Schools Safe in Earthquakes*, OECD, Paris

Penelis, G 2001. *Pre-Earthquake Assessment of Public Buildings in Greece*, International Workshop on Seismic Assessment and Rehabilitation of Structures, Athens Greece, Jan, 2001.

Sachs, J 2005a. The End of Poverty: How We Can Make It Happen in Our Lifetime Penguin

Sachs, J (ed.) 2005b. Investing in Development: A Practical Plan to Achieve the Millennium Development Goals, Report of the Millennium Project, Earthscan

Schilderman, T 2004. "Adapting traditional shelter for disaster mitigation and reconstruction: experiences with community-based approaches", *Building Research and Information*, 32(5), 414–426

Spence, R, Lopes, M, Bisch, P, Plumier, A and Dolce, M 2005. *Earthquake Risk Reduction in the European Union*, European Association for Earthquake Engineering, (unpublished document).

Spence R 2007. "Saving lives in earthquakes: successes and failure in seismic protection since 1960", *Bulletin of Earthquake Engineering*, 5139–5251.

Spence, R, So, E, Cultrera, G, Ansal, A, Pitilakis, K, Campos Costa, A, Tonuk, G, Argyroudis, S, Kadderi, K and Sousa, M 2008. Earthquake Loss Estimation and Mitigation in Europe: a review and Comparison of Alternative Approaches, 14th World Conference on Earthquake Engineering, Beijing.

So, E, Spence, R, Khan, A, and Lindawati, T 2008. Building Damage and Casualities in Recent Earthquakes and Tsunamis in Asia: A Cross-Event Survey of Survivors, 14th World Conference on Earthquake Engineering, Beijing.

Tucker B 2004. At the Turning Point for Global Earthquake Safety, Geohazards International, www.geohaz.org

WHO 2000. *WHO Report on Global Surveillance of Epidemic-prone Infectious Diseases, Chapter 4: Cholera*, World Health Organisation, WHO/CDS/CSR/ISR/2000.1

Wisner, B 2004. "Swords, plowshares earthquakes floods and storms in an unstable globalising world", *J Nat Disaster Sci*, 26(2), 63–72.

Yelland, R and Tucker, B (eds.), 2004. *Keeping Schools Safe in Earthquakes*, OECD, Paris.

Seismic Protection of Monuments

T.P. Tassios

Abstract Seismic protection of buildings is a well established scientific and technological domain. The knowledge needed to this end, is not the subject of this lecture. Instead, only the particularities of the seismic re-design of *monuments* are discussed, resulting in their structural repair or strengthening.

The main particularity of this endeavour is that a monument, besides its possible practical use and its economical value, requires a lot of other values to be respected during its aseismic retrofitting, such as its aesthetic Form, the authenticity of its materials etc. In order to respect these values, the engineer tends to minimize structural intervention – thus, violating social values such as the preservation of the monument for future generations, the protection of human lifes, etc. An optimization is needed, and the lecture attempts to describe the necessary procedures to this end. Besides, emphasis is given to the particular difficulties in the determination of the resistance of masonry, as well as in the selection of suitable methods of Analysis, taking into account the specificity of each monument.

1 The Significance of the Subject

It is broadly accepted that in seismicly prone regions, the seismic behaviour of monuments is of a paramount importance.

First, because of the cultural need to maintain and transfer these monuments to future generations. To this end, more or less drastic structural interventions (repair or strengthening) are implemented[1], with the lowest possible consequences on the "monumentic" values.

T.P. Tassios (✉)
National Technical University of Athens, Athens, Greece
e-mail: tassiost@central.ntua.gr

[1] In this respect, specialists do not anymore share the view that "since a Monument has withstood previous earthquakes, it will continue to resist any future seismic action".

A.T. Tankut (ed.), *Earthquakes and Tsunamis,* Geotechnical, Geological, and Earthquake Engineering 11, DOI 10.1007/978-90-481-2399-5_5,

Second, our interest for the seismic resistance of monuments is also encouraged by the legal obligation to protect human life (of the neighbours, curators, visitors or even inhabitants of the monumental building)

However, the preparation of design documents regarding structural interventions of monuments is frequently facing several *difficulties*, not encountered in the case of non-monumental buildings, such as:

- Additional uncertainties related to the available resistances of building components
- Particularities in selecting the appropriate method of Analysis, suitable (i) to a given typology of the monument and (ii) to the level of resistance uncertainties
- Difficulties in selecting
  - An appropriate design value of seismic action, such that the respective necessary intervention will not jeopardise the monumentic values of the monument, and
  - Appropriate techniques with an optimum level of reversibility / re-interventionality.

Because of these difficulties, the approval of submitted design – documents is frequently an occasion for controversial discussions between engineers, on the one hand, and architects and archeologists, on the other.

2 Structural Interventions and the Conflict of Values

These controversies are but a reflection of the contrarieties between the following "Principles" (Values and Requirements) related to the structural interventions in monuments:

a) Monumentic[2] Values

a1: Form (aesthetic value).
a2: History (symbolic value).
a3: Preservation of ancient building-Techniques and Materials (technical value).

b) Social Values

b1: Preservation of the cultural Memory of a Monument (integrity, survival).
b2: Adequate safety against normal actions and Earthquakes (value of human life).
b3: Modern use of Monuments.
b4: Cost-reduction of the structural intervention.

[2] This neologism is a very useful term to express concepts and things related to Monuments.

c) Performance – Requirements regarding the structural intervention (Technical Values)

c1: Reversibility level and or re-interventionality level
c2: Durability
c3: Technical reliability

Every intervention aiming at a structural repair or strengthening of a Monument entails some inevitable *harm* to several of these values and performance-requirements, depending on the actual condition of the Monument and the available technologies. It suffices perhaps to give a few typical examples:

- In order to preserve the Form and the integrity of a seismicly vulnerable Monument, a rather costly strengthening solution is adopted, consisting in (i) extended grouting of masonry walls and (ii) change of the old (completely decayed) roof. Thus, the following "principles" were violated: a3, b4 and c1.
- A second typical example may be the case of a monumental building made of precious historical materials to be completely preserved; the solution here was to offer seismic safety by means of external buttresses. The violated "principles" here were: a1 and a2.
- In a third case, to avoid any harm to monumentic Values of a delicate Monument, some rather simple and provisional structural interventions were decided, offering a seismic resistance lower than the one required for modern important buildings. In this case, a remarkable violation of the social value of "human life safety" (principle b2) was accepted, together with a transgression of the durability requirement (c2).

Apparently, in all these cases, Authorities have sought an optimization[3] of Principles, and came to their final decision, knowingly of the partial violation of the "set of Principles". In this respect, it is reminded that such an optimization cannot be reached by means of just scientific judgments: The values entering the game are of different nature; they are *not* amenable to identical "units" – they are not quantitatively comparable between each other!

That is why, in the field of structural interventions of Monuments, only managerial (almost political) decisions are feasible; weighing factors for each of these Principles may be (directly or indirectly) discussed within an interdisciplinary group, and a final "optimal" decision be made.

Such an optimization process, directly affects some important technical issues related to the seismic (re)design of Monuments: The design value of seismic actions has to be decided, taking (also) into account its eventual consequences on monumentic values, too, as well as on costs and technical performances. Thus, a sort

[3] See Sections 9 and 10.

of "negotiation"[4] of design seismic actions is initiated: Disproportionately high design-values, serve the "human life" and the "integrity" principles, but they may jeopardise some monumentic values and performance requirements. Therefore, a better overall intervention (an "optimal" solution) may be sought, based on possibly lower design-values of seismic actions, i.e. on higher exceedance probability. The same holds true for the selected intervention schemes and technologies; they should also be finally decided following a similar optimization process.

In order to facilitate such a decision making process, further rationalization of data is needed regarding the "Importance" of a given Monument, as well as its "Visitability" level.

3 Importance, Visitability and Acceptable Damage-Levels

- Seismic actions' values for the re-design of Monuments may also depend on acceptable damage-levels, which in their turn will be decided on the basis of the importance of each Monument. That is why in many Countries, a categorisation of Monuments is available to designers, as follow:

  I1: Monuments of universal importance
  I2: Monuments of national importance
  I3: Monuments of local interest

- Another useful tool towards a rationalisation of decision making regarding structural interventions, is the categorization of the occupancy of Monuments: Higher occupancy means higher concern for human lives against earthquakes, and therefore higher seismic actions' design values. That is why engineering decisions would be facilitated if a "visitability" categorisation of Monuments would be made available, such as in the following list:

  V1: Almost continuous presence of public or frequent presence of large groups

  - Inhabited building in historical city centres
  - Monuments used as Museums
  - Monuments continuously used for worshipping

  V2: Occasional habitation or intermittent presence of small groups

  - Monuments visited only under specific conditions
  - Remote and rarely visited Monuments

[4] Afterall, design seismic actions regarding modern buildings are *also* negociated: The socially acceptable "probability of exceedance" of seismic actions imposed by actual Codes, depends on several variables, such as the actual economical level of the Country and the social importance of the building, i.e. on non-scientific data. The difference with Monuments is that in this case, such a "negociation" is taking place within a broader multiparametric space, including many additional Values and Requirements.

V3: Entrance allowed only to Service-Personnel. Visitors stand only outside the Monument.

- Combining the aforementioned Importance-levels and visitability-levels, it is possible to decide the acceptable Damage-levels ("I" for negligible damage, up to IV for serious damage).

Such a possible matrice is given here below (indicatively though):

Acceptable damage-levels (I–IV) under the design seismic actions		Prevailing values: Human life and monument's integrity			Prevailing values: Form and history
Visitability		V1	V2	V3	
Importance level	I1	I	II	II	I
	I2	I	II	II	II
	I3	II	III	III	III

It is believed that such an approach may substantially facilitate rational decision making, related to structural interventions of Monuments against seismic actions.

4 Documentation and Uncertainty Levels

a) Long experience shows that the structural design document regarding seismic strengthening of a Monument is an integral part of the broader study of the Monument; history and architecture of the Monument are indispensable prerequisites for the Structural Design, in order to account for all initial and consecutive construction *phases*, previous repairs etc.
b) Description of existing and or repaired *damages* (visible or possibly hidden ones), together with their in-time evolution. Even short term monitoring may be helpful.
c) Systematic description of the in-situ *materials*, including their interconnections – especially in the case of three leaf masonry walls. Connections of perpendicular walls are thoroughly investigated.
d) Results of *experimental* investigations regarding:

- geometrical data,
- internal structure,
- in-situ strength of materials,
- structural properties of masonry walls,
- dynamic response of building elements,
- subterranean data,

as well as results of possible previous *monitoring* installations (displacements, settlements, internal forces, humidity, groundwater level, cracks' opening, seismic accelerations, environmental data etc).

d) Description of the *structural system*
e) Description of the soil and the foundation

The combination of all these data is an indispensable basis for the structural re-design of a Monument, since the structure of a Monument is a rather silent black-box; every effort is justified to make this box to talk. However, a lot of uncertainties will remain. And our scientific duty is to be conscious of these uncertainties and of the way they may affect Analysis (hidden discontinuities, behaviour of connections etc), as well as Resistance determination (hidden voids, weathered or inhomogenious materials, etc). Otherwise, our computational efforts may not be able to correctly evaluate the seismic resistance of the Monument and to appropriately design its best strengthening.

Long experience shows that for each primary structural member of the Monument, an appropriate level of reliability of documentation should be assigned, referring separately to basic data, such as:

- Dimensions, eveness and verticality
- Composition transversally to the element (e.g. three leaf masonry?)
- Connections with neighbouring elements.
- Strengths of constitutive materials, etc.

For each of these categories of data, a "Documentation-Reliability level" should be assessed in each particular case, (indicatively: "missing", "inadequate", "sufficient"). In accordance with this characterization, Analysis and Resistance evaluation will be effectuated more accurately (see Sections 5 and 6).

5 Structural Analysis

For given external or internal actions, the action-effects (stress resultants) on all critical cross sections of the structural members are determined by means of an appropriate method of Analysis (static or dynamic, linear or non-linear, on discrete structural elements or on finite elements). The following comments may be useful in selecting and applying such methods.

a) First, the *structural system* has to be clearly identified; the rich data included in the Documentation of the Monument are very helpful in this respect.
b) *Selection criteria* of the method of seismic Analysis, appropriate for the Monument under consideration:

   - For more important and or complicated Monuments, more sophisticated methods appear more justified.

- For each of the main typological categories of stone monuments, some particular methods may be more appropriate. Indicatively:

  - For normal masonry walls: Force path method, Finite elements method (FEM), Non linear static method (pushover), etc.
  - For slender structures: Linear dynamic method.
  - For arched structure: Force path method, Kinematic plastic method, etc.
  - For "sculptured stones/dry joints" Monuments (greco-roman typology): Friction and Rocking Response method.

- For each acceptable *damage-level* (Section 3), a respective method of Analysis may be used. Indicatively:

  - In case of targeted negligible damage under the design earthquake, only linear Analysis is applicable.
  - In case of extensive accepted damages, a non-linear static Analysis seems to be more suitable.

- The available level of "Reliability of the Documentation" (Section 4) should also be taken into account when selecting the appropriate method of Analysis. Low RD-levels are not compatible with highly sophisticated methods.

c) A warning may be useful, related to the use of data found by means of *dynamic experimental excitations*. These data, applicable for structural identification, should not be used for spectral response calculations; in fact, eigen frequencies determined by means of dynamic excitations, are considerably higher than real frequencies under actual displacement and damage conditions.
d) It is suggested to check if the chosen method of Analysis is able to "reproduce" roughly (through computation) previous damages the monument has suffered because of previous loading conditions (mainly previous documented earthquakes).
e) In the case of linear Analysis, the use of a "behavior factor" q (reduction factor) for Monuments is a rather delicate subject. In the case of traditional masonry structures, values as low as 1,00 to 1,50 are recommended for unconfined masonry, or 1,50 to 2,00 for fully confined walls in a well tied monumental building – provided however that at least a medium damage-level is acceptable.

6 Evaluation of Resistances

One of the most remarkable particularities of the structural re-design of Monuments is our difficulty to determine the resistances of critical regions of stone building-elements, with an accuracy comparable to the precision of the determination of action-effects. Thus, the inequality of safety ($E \ngtr R$) becomes a rather loose condition, leading to overconservatist or to risky solutions.

This unpleasant situation is partly due to the fact that research financement and research glamour is normally offered to subjects related to Analysis, rather that Resistance determination...

In the field of structural assessment and re-design of Monuments, it is of a paramount importance to overcome this *scientific weakness*; thus, every effort is justified in order to better evaluate available Resistances of critical regions of "masonry" walls, or "sculptured stones/dry joints" building elements. To this end, empirical formulas or simple rules of thumb do not serve the purpose alone: Detailed preliminary inspection and in-situ experimental investigation will be the indispensable basis for any subsequent computational step of resistance determination.

6.1 Semi-Analytic Resistance Determination

a) The following *data* are first needed in order to evaluate the basic resistance characteristic of a masonry, i.e. its compression strength perpendicularly to its layer:

   - Photographic view of the face of each critical region of the wall or pillar.
   - Calculation of mean values of length (l_{bm}) and height (h_{bm}) of the blocks, as well as the mean value of thickness of joints (t_{jm}).
   - By means of dense vertical sections, a nominal index "i" of blocks' interlocking is calculated: Within a representative height (H_m) of masonry, such an index may be defined in the most unfavourable section as $i_{min} = \Sigma U_b : H_m$, $(0,1 \lesssim i_{min} \lesssim 0,9)$ where "U_b" denotes the heights of blocks cut by the vertical section in consideration.
   - Examination of the structural composition of the wall perpendicularly to its plane. To this end, direct observation through temporarily opened holes, or televisioned endoscopy, or even georadar (and or sonic tomography) may be used, in order to find out if the wall may be considered as one-leaf or two (or three)-leaf masonry. In Monuments of lower importance, some transverse core-takings may also be helpful in this respect.
   - Strength and deformability of blocks and mortars are measured in situ (by means of appropriate ND methods), or in-lab (on appropriate representative samples).

b) It is now possible to use a well established empirical expression predicting the vertical compression strength of masonry, under the condition that, among its parameters all the physical variables shaping the overall compression strength are included, such as:

   - nature of blocks (stones, bricks)
   - roughness of blocks' surface
   - strengths and deformability modulus of blocks and mortars
   - average joint-thickness (and its normalized value $\alpha = t_{jm}:h_{bm}$)
   - interlocking index i_{min}

– transversal structural composition of the wall (one or three-leaf masonry).

Under these basic conditions, appropriate empirical formulae will be used, in order to predict masonry compression strength (f_{wc}), separately for one-leaf and three-leaf masonry.

c) *Tensile* strength (f_{wt}) should be carefully estimated on the basis of appropriate expressions, accounting for the favourable contribution of friction resistance of blocks in case of weak mortar. Otherwise, tensile strength may be estimated as a variable percentage (1:3 up to 1:15) of compression strength, depending on the magnitude of compression strength.
d) *Diagonal compression* strength (f_{wi}) may be estimated as a function of f_{wc}, f_{wt} and the transversal tensile stress (σ_t).

6.2 Additional In-Situ Strength Determination

In some Monuments, it may be allowed to complement the data mentioned in Section 6.1, by means of in-situ strength determinations. To this end, flat-jacks are inserted into appropriately deep horizontal slots, so that an in-situ masonry "specimen" to be shaped and tested.

6.3 In-Lab Testing of Replicas

In the case of Monuments of higher importance, provided that the experimental investigations described in Section 6.1.a have offered sufficient information, it is suggested to reproduce in laboratory (piece by piece) a part of the critical region of the masonry under consideration. Thus, specimens will be available – replicas of the actual wall, to be subsequently tested under vertical and diagonal compression. The dimensions of these replicas should be large enough, for the failure mechanism to be freely developed.

6.4 Resistance of Walls or Columns Made of Sculptured Stones/Dry Joints (Greco-Roman)

Walls: Their compression strength under distributed load is influenced by possible internal and interface irregularities of some blocks, but it is generally of the same order as the strength of individual stones. Resistance against local load may be predicted via the provisions of Mechanics.

Columns made of stone drums: Two resistance mechanisms should be documented in advance:

- Friction: (i) The relationship of friction coefficient and normal stress should be well known and a statistical dependence has to be established, for several humidity conditions of the joint. (ii) The constitutive low "imposed slip/mobilized friction resistance" should be experimentally found, for the specific conditions of the drums in consideration. To this end, in-lab tests on representative replicas are needed.
- Strength of a block or a drum in rocking position: Relevant in-lab tests on replicas may be needed for the determination of this strength (inclined compressive force, concentrated on a free edge), unless a reliable analytic model might be satisfactory.

7 Assessment of the Actual Seismic Capacity of the Monument

To this end, normally simple static linear models of Analysis are first used, up to a value of seismic actions revealing a local insufficiency. Subsequently, depending on the acceptable damage level of the Monument (Section 3), the assessment may end here, or for Monuments of relatively lower importance, further increase of seismic actions may be considered by using more sophisticated methods of Analysis, as described in Section 5, up to the level of acceptable damage.

However, in the case of slender Monuments, the use of a simple dynamic method of Analysis is needed since the very beginning of the assessment.

8 Selection of the Re-Design Earthquake for the Structural Intervention

This issue is of fundamental importance. In Section 2 we faced multiple contrarieties: our wish to ensure a high seismic resistance for a Monument may necessitate structural interventions as "heavy" as to harm some monumentic or technical Values. Thus, we have recognised the need for an "optimization" or, in other words, the need to negociate the re-design earthquake to be finally used in our calculations. This chapter attempts to summarise the procedures to be followed in order to select an "optimal" value of seismic actions.

8.1 Historic Data

The behavior of the Monument under specific earthquakes of the past, is well studied and described in the Documentation (Section 4a, b). Although it is not easy to quantify the intensity of those earthquakes, historical data do frequently offer helpful collateral information, such as: behavior of normal buildings in adjacent areas, fall of objects etc. Sometimes, in case of very large seismic events, Archeoseismology is able to give us some further quantified data.

On the other hand, we may also derive quantitative information about the intensity of previous earthquakes, based on the behaviour of pre-existing repairs or strengthenings.

To the extent such a methodology is feasible, it will be possible to have a first estimate of the overall seismic capacity of the Monument as it stands. This knowledge may assist us in selecting a couple of seismic load values for the re-design of our planned structural interventions.

8.2 Actual Seismological Studies

This is the most "easier" source of information; but, as it is well known, the expected seismic load values are related (i) to several probabilities of exceedance, and (ii) to a conventional life-time of the building. Thus, this scientifically established knowledge has not an "absolute" validity: First the life-time of a Monument cannot be clearly expressed; by definition, "monuments" should live "for ever". And second, the notion of an acceptable probability of exceedance is not too friendly to Curators of Monuments.

However, the designer may face these two issues as follows.

(i) Take into account first the probability actually accepted for existing old urban buildings, and second the probability specified by the Code in force for buildings of very high social importance.
(ii) In place of a life-time duration, one may consider a reasonable period for radical repair and strengthening of the Monument in consideration.

In any event, the seismic load values dictated by actual seismological studies, are but one of the candidate-values for the final re-design; within the admitted "negociation" for optimization, this seismological input loses its character of undeniability.

8.3 Selected Seismic Loads

Based on these two sources of information, as well as on his/her own experience, the designer selects two (or three) temporary values of seismic load to be submitted to the final optimisation needed (Section 10). This "seismic load" may be expressed in appropriate terms (depending on the importance and the structural specificity of the Monument), such as ground acceleration, accelerograms and the like.

In any event, none of these seismic load values could be lower than a minimum level commonly acceptable.

Note: Thanks to the concept of "levels of acceptable damage" during the re-design earthquake, lower seismic load values have more chance to be satisfactory.

9 Preliminary Designs

The fundamental process of the aforementioned optimisation, needs to be fed by at least two alternative solutions of structural intervention. Assuming that the repair or strengthening techniques were correctly chosen, it remains to check the favourable and adverse consequences of two seismic load design values. In what follows, the pertinent procedures are summarised.

a) As it is well known, there are several available intervention philosophies, such as:

   - Local repair or strengthening of a building member, (or its connections to adjacent elements), without external modification of its form.
   - Installation of ties and confinements.
   - Addition of new building elements (permanent or provisional) in order to ensure the wished global seismic resistance to the Monument. In this category of structural measures belongs also the case of adding or strengthening a diaphragm (for a better distribution of seismic forces), as well as the addition of hidden and completely reversible dumping devices.
   - Possible construction of seismic isolation underneath foundations, (although such a technique may be applicable only in relatively recent Monuments).
   - Monolithic transportation of the Monument to a more favourable place – a solution suitable only in relatively small size Monuments.

   Any intervention technique to be used, has to be appropriably checked against all Principles described in Section 2, especially those related to the specific performance requirements (Section 2c); in this connection too, there is space for another category of optimisation – which however is not further discussed in this paper.

b) Thus, assuming that the proposed intervention techniques were correctly chosen, at least two (or better three) seismic load values have to be used for the redesign, as they were preliminarily adopted in Section 8.3. For these seismic loads, the respective structural interventions are designed, and preferably illustrated in preliminary drawings.

   The consequences of such interventions on all Principles enumerated in Section 2 will be now considered in details, to be clear enough for the interdisciplinary decision-making Group to deliberate as in the following paragraph.

10 Selection of the Final Optimal Solution

In-view of the interdisciplinary nature of Principles involved in the decision-making process, it is suggested that the final selection of the structural intervention scheme of seismic strengthening of the Monument, be effectuated by a representative Group – including the Owner of the Monument. (However, depending on available local habits, decision may be taken by just one responsible agent of the Authorities).

In any event, the process for the final judgement may be conceptually described as follows:

a) For each of the Principles enumerated in Section 2, a scale of "level of satisfaction", graded $G = 1$ to $G = 5$ (from low to high degree of satisfaction) is established.
b) For the specific Monument under consideration, a lowest acceptable level ($G_{i,min}$) of each of these ten Values (i) is decided.
c) For the same Monument, weighing factors "f_i" are agreed, expressing the relative importance of each of the ten Values (with $\sum f_i = 1$). Obviously, it is expected that the highest values will be given to the factors regarding Form and Human life.
d) An evaluation is needed of the way a given intervention solution affects the aforementioned Values; grades G_i ($i = 1$ to 10) are agreed respectively.
e) The level of aseismic resistance (R) ensured to the Monument by the solution envisaged, was found in Section 9b. It is reminded however that $R \not< R_{min}$, a commonly acceptable minimum level of seismic capacity.
f) An estimator of overall efficiency of the proposed solution is calculated

$$e = \frac{R}{R_o} \cdot \sum_{i=1}^{10} f_i \cdot (G_i - G_{i,\min})$$

where R_o is an arbitrary, sufficiently high resistance-level (e.g. that wished for an extremely important modern building).

The solution with the higher score may be selected, unless (as it is very common) further *refinements* are requested by the Authorities.

It has to be noted that the whole procedure is also applicable without its quantifiable version: Experts may offer their opinion, based on their qualitative judgement after a thorough examination of the consequences of each solution. Afterall, the main interest of this procedure is that it may act as an open reminder of the multifold aspects of the design for the aseismic protection of Monuments.

Note: Because of the descriptive and qualitative nature of this lecture, no References are given here.

Mitigation of Seismic Risk in Italy Following the 2002 S.Giuliano Earthquake

Mauro Dolce

Abstract In order to mitigate seismic risk, a full range of actions, from prevention to emergency and post-emergency management, from awareness campaigns to preparedness must be undertaken. These actions are usually carried out by several subjects, among which the Civil Protection authorities are the most involved ones and in the entire cycle of activities. From an engineering point of view this implies several technical activities, some of which are quite peculiar. Very often, actions for seismic risk prevention are undertaken after destructive earthquakes that have a great impact on public opinion. This is the case of the S.Giuliano Earthquake of October 31st, 2002, which was not a big earthquake but was quite shocking, because of the death of 27 children in their collapsed school. In the paper the full range of actions undertaken by the Italian Civil Protection aimed at mitigating seismic risk after the S.Giuliano earthquake are described.

Keywords Seismic risk · Prevention · Emergency · Preparedness · Civil protection · Seismic code

1 Introduction

The Italian territory is certainly subjected to a high seismic hazard, although it is not the most prone area in Europe, if one looks at the situation of Greece and the western part of Turkey (see Fig. 1). However, due to the high population density and the high vulnerability of the buildings, Italy is probably the country with the highest seismic risk. As a matter of fact, some figures related to the far and recent earthquake history in Italy are significant:

- Since 1000 A.D. nearly 30,000 events occurred (220 of them with MCS intensity $\geq$ VIII)

M. Dolce (✉)
Civil Protection Department, General Director of Seismic Risk Office, Rome, Italy
e-mail: mauro.dolce@protezionecivile.it

A.T. Tankut (ed.), *Earthquakes and Tsunamis,* Geotechnical, Geological, and Earthquake Engineering 11, DOI 10.1007/978-90-481-2399-5_6,

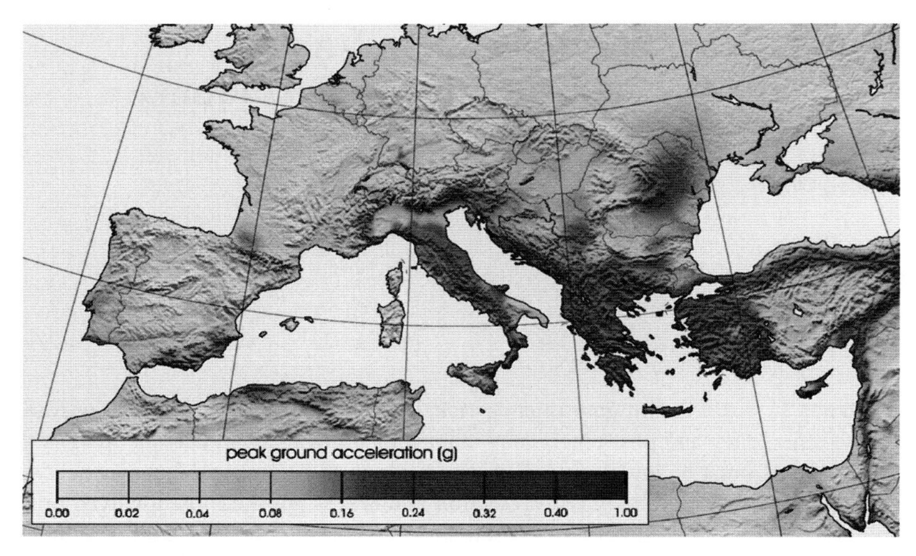

Fig. 1 European-mediteranean seismic hazard map (Jiménez et al. 2003)

- In the last two centuries earthquakes caused about 150,000 casualties; moreover, they damaged and/or destroyed a great part of the Italian historical and cultural heritage, whose value is not quantifiable
- In the last 40 years earthquakes caused monetary losses for over 130 billions Euros (see Table 1)

Table 1 Costs of recent Italian earthquakes

Earthquake	Year	Magnitude	Epicentral intensity MCS	Casualties	Costs (m*e*) (2005 updated)
BELICE	1968	6.4	IX–X	175	14,515
TUSCANIA	1971	4.5	VIII–IX	0	438
ANCONA	1972	5.2	VIII–IX	0	5,884
FRIULI	1976	6.4	IX–X	989	36,629
CALABRIA E SICILIA	1978	5.3	VIII	0	390
VALNERINA	1979	5.9	VIII–IX	9	1,062
IRPINIA	1980	6.9	X	2°914	53,637
MAZARA DEL VALLO	1981	4.6	VI	0	382
POZZUOLI	1983	4.0	VII–VIII	0	1,062
LAZIO-ABRUZZO	1984	5.2	VII	3	2,695
SICILIA	1990	4.7	VII–VIII	17	2,563
EMILIA-ROMAGNA	1996	4.7	VIII	0	136
MARCHE-UMBRIA	1997	5.9	IX–X	11	11,009
POLLINO	1998	5.6	VII	1	667
EMILIA-ROMAGNA	2000	4.4	VI	0	29
			TOTAL		131,097.34

In Table 1 there are reported the expenses relevant to the Italian earthquakes since 1968. The most recent earthquakes are not included, as only a part of the required expenses have been carried out. Table 1 emphasises the disproportion between the actual costs and the potential destructiveness of earthquakes, expressed through their magnitude, although some considerations should be made on the reconstruction policy adopted after recent earthquakes. As a matter of fact, the recorded expenses were not aimed at purely repairing structures, but also to realise their seismic upgrading. Moreover some general investment for soil stabilisation, infrastructure improvement and industrial activities promotion were carried out, in order to speed up the socio-economical recovery or development of the less-favoured areas stricken by earthquakes.

The situation has not improved in the last years, since the turn over of the construction is very slow in Italy and most of the vulnerability problems that caused the high costs of past earthquakes have been overcome only in a negligible extent. This appears very clear from the projections based on last two century history of earthquakes. As a matter of fact, when applying the scenario simulation program that is currently used in the Civil Protection Department (NCPD) to all the earthquakes of magnitude greater than 5 that occurred in the XIX centuryor in the XX

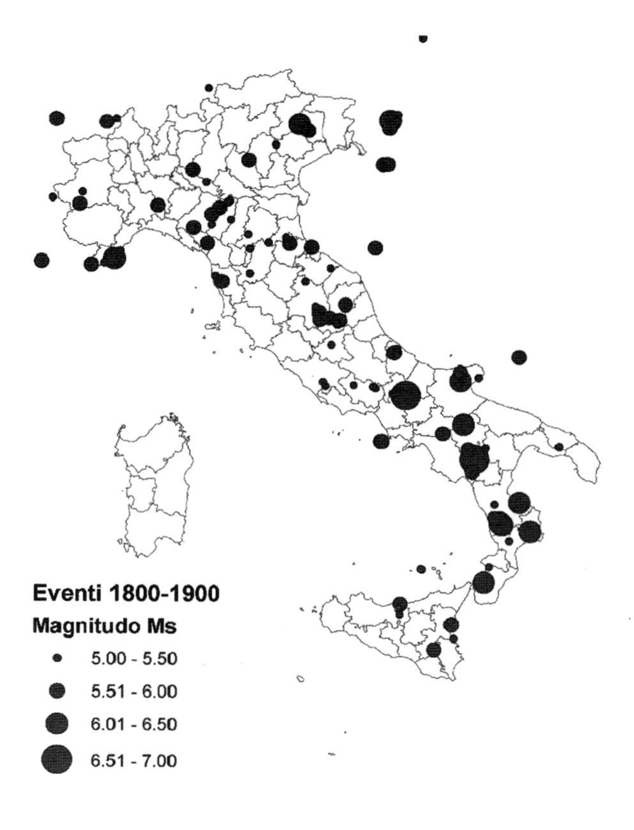

Fig. 2 XIX century earthquakes M > 5

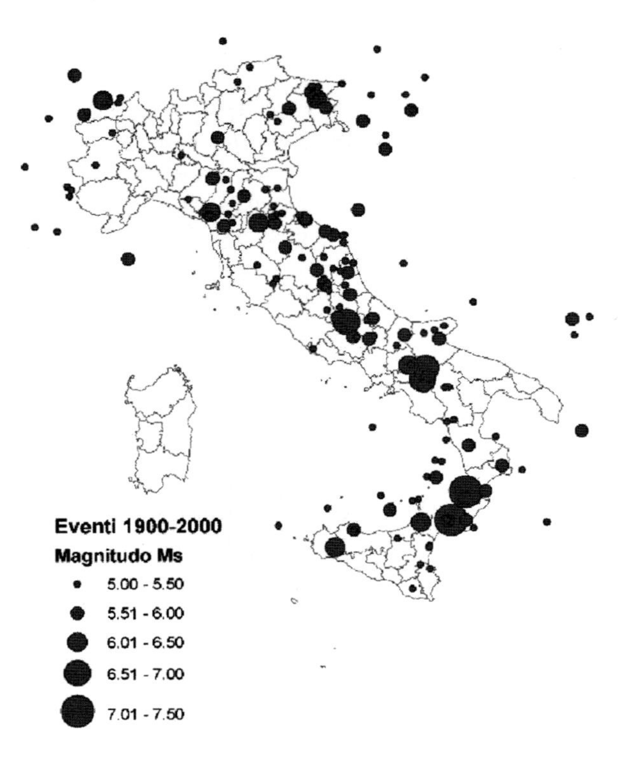

Fig. 3 XX century earthquakes M > 5

century (shown in Figs. 2 and 3) to the current Italian housing inventory (national population census of 2001), the following figures are found:

500–2000 dead or injured people / year
→ 50,000–200,000 in the XXI century
1–2 billions / year
→ 100–200 billions in the XXI century

These cost estimates are relevant to houses only. Overall costs should include also public and office buildings, monumental buildings, factories and infrastructure. Possible increases of the order of 50–100% can be expected.

The impact of earthquakes in Italy is undoubtedly very high, much more than expected in other areas with high seismic hazard. The reasons for such high earthquake impact ad costs is essentially related to the high vulnerability of the Italian real estate, because of:

- the large number of historical and old buildings,
- the degradation of many urban suburbs,

- the large number of illegal buildings, particularly in Southern Italy, where the seismic hazard is higher,
- the bad knowledge of the seismic hazard in the past,
- the inadequacy of technical standards and of their application.

On the other hand, it is well known, that without any seismic emergency, resources are not usually allocated to implement policies and strategies for seismic risk mitigation. This is the main reason why significant innovation and action programs for seismic risk reduction are usually introduced just after destructive earthquakes, when society is devastated and the risk is highly perceived.

Though having developed a significant experience in earthquake recovery, Italy did not go beyond this approach:

- Significant improvements in seismic code and seismic zonation were implemented only after destructive earthquakes
- Direct activities for reduction of seismic vulnerability were implemented in areas affected by recent strong earthquakes.

A clear demonstration of this attitude is given by the history of the classification and of the seismic code in Italy.

Before Italy unification some codes were enforced in some independent states for post-earthquake reconstruction. The same occurred in the Italian State after 1861, according to the following examples:

1783 → a code was issued by the King of the two Sicilies (Southern Italy Kingdom), for the reconstruction after a destructive sequence in Calabria,
1859 → another code was issued by the Pope (Central Italy) after the Norcia earthquake,
1908 → after the Messina earthquake (83,000 casualties, Ms = 7.3) a seismic zonation was introduced by the Italian Government, and a building seismic code was issued,
1915 → after the Avezzano earthquake (30,000 casualties, Ms = 7.0): many areas of Central Italy were classified as zone II (S.C. = 0.07),
1980 → after the Irpinia-Basilicata Earthquake (about 3,000 casualties, Ms = 6.9): a new zonation, finally based on a consistent probabilistic approach was enforced.

This approach is exemplified by the seismic zonation maps shown in Fig. 4. The seismic zonation introduced after the 1908 Earthquake was derived from the occurrence of that earthquake and some others that struck many sites in Calabria in the previous decade. Until 1980 only the areas that were affected by destructive earthquake during the XX century were included in the seismic zonation, according to two different grades (1st and 2nd category), so that in 1980, when the Irpinia earthquake occurred, only 25% of the Italian territory was included in the seismic zonation. From 1981 to 1984 a new zonation based on a consistent probabilistic approach and on some centuries of seismic history was progressively enforced, so

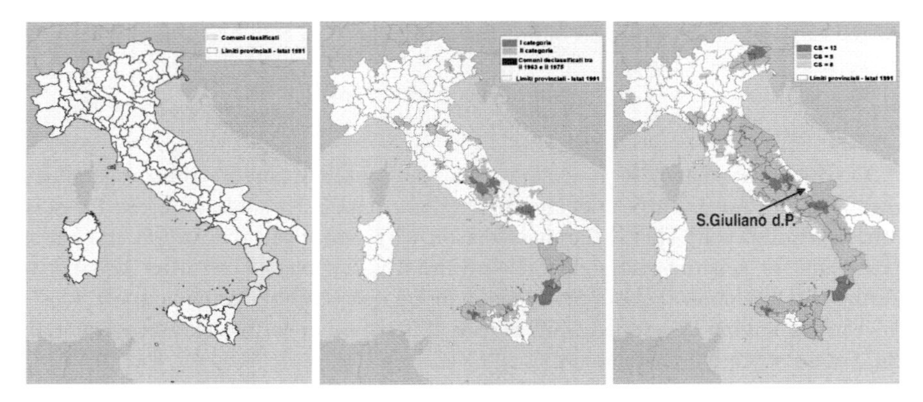

Fig. 4 Historical evolution of seismic classification in Italy (De Marco and Martini 2001) (See also Color Plate 5 on page 211)

that in 1984 about 45% of the Italian territory was included in the seismic zonation (1st, 2nd and 3rd categories).

However, the Italian seismic code remained unnecessary misleading and unclear in its basic scopes. Moreover a behaviour factor of about 5 was implicitly assumed for most structural types, without enforcing adequate detailing. The ductility concept was not explicitly reported in the code. Few information was given about assessing, upgrading or retrofitting existing buildings.

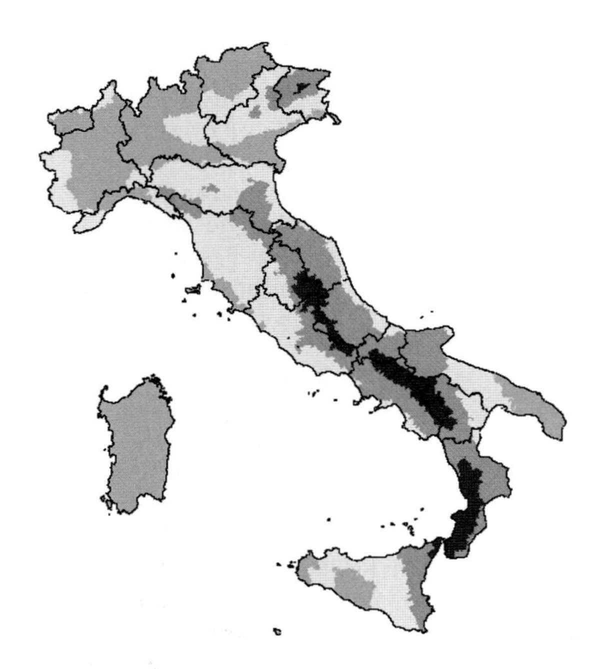

Fig. 5 Seismic classification proposal of 1998 (Gavarini et al. 1999) (See also Color Plate 6 on page 211)

Nevertheless, several seismic code and seismic zonation modifications were proposed in Italy in the recent past by the scientific community, but they were not enforced, probably because, fortunately, no destructive earthquake had occurred in the meanwhile. Among them, a new seismic code proposal in the early 80's and a new seismic zonation in 1998 (see Fig. 5).

A significant change has recently occurred only due to the strong impact that the October 31st, 2002 S.Giuliano earthquake, causing the death of 27 children in their school, had on the public opinion.

In this paper the lessons learnt specifically from the S.Giuliano Earthquake and the measures taken by the Italian Government, and by NCPD in particular, are described and commented.

2 The S.Giuliano Earthquake and the Relevant Learnt Lessons

On October 31st and November 1st, 2002, two earthquakes of magnitude 5.4 and 5.3 hit the area at the border between Molise and Puglia in Southern Italy. The damage pattern shown in Fig. 6 qualified the earthquake as intensity VII MCS (Mercalli-Cancani-Sieberg scale), with one notable exception: the village of San Giuliano di Puglia. S. Giuliano suffered damage corresponding to macroseismic intensity I = VIII–IX MCS, causing 30 fatalities, including 27 children due to the collapse of their primary school.

Inside the same San Giuliano town the damage was varying from moderate to severe, with some total collapses (including unfortunately the primary school) concentrated in a restricted area (Goretti and Dolce 2004).

The towns surrounding San Giuliano also suffered varying degrees of damage, with lower felt intensities (I = V–VI to VII MCS), no fatalities, and relatively few injuries

The public opinion was shocked by the death of the 27 children and pushed the government to take the needed countermeasures in order to mitigate the seismic risk in Italy. The Civil Protection Department contributed substantially in order to activate a process with the following steps, taking into account the lesson learnt from the S.Giuliano and other recent earthquakes:

- Transferring SoA knowledge into updated seismic zonation and national norms;
- Allocating adequate funds for seismic risk mitigation, to make seismic safety assessment and rehabilitation of existing constructions;
- Transferring SoA knowledge to engineers, professionals and operators;
- Re-addressing and funding research activities to get products which can immediately and robustly be exploited in the seismic mitigation process;
- Increasing the awareness and the culture of civil protection in the population.

In order to understand the actions undertaken it is useful to remind some lessons learnt from the S.Giuliano Earthquake.

Fig. 6 Typical damage of masonry buildings in S.Giuliano di Pugliia

1. In the affected area earthquakes occurred several centuries ago. In 1456 earthquake damage was reported in many villages. Most severe effects were reported during the 1627 earthquake. In 1805 seismic intensity was at the damage onset. No destructive earthquake occurred in the area. Lesson learnt is to beware of areas with rare events. People forget easily the consequences of seismic events, making it difficult to develop any local seismic culture in the area.

2. The historical centre of S.Giuliano was built on rock, while the expansion area, where most of the damage occurred, was built on soft clay soil. Lesson learnt is that site effects can play a critical role in transforming a non-destructive earthquake in a devastating event and that an incorrect land use can contribute to the disaster. No destructive earthquake was reported in the past probably because villages were, originally, built on rock.
3. In the classification proposal of 1998 the affected area was classified as moderate seismicity area. Moreover the Italian building code was obsolete. Lesson learnt is that authorities should update seismic classification and code to the latest available information, once a sufficiently wide scientific consensus is reached.
4. Due to the obsolescence of both seismic code and seismic zonation, buildings constructed in the past are not safe enough and are not so safe as they could have been. Lesson learnt is that there is an urgent need of seismic assessment of the buildings and infrastructures, at least those having a crucial role in emergency or implying relevant consequences in case of collapse.
5. Even in the areas were the felt intensity was 7 or less, considerable damage was suffered by the cultural heritage, especially by churches, once again emphasizing their high seismic vulnerability and the high risk deriving even from low intensity events. Lesson learnt is that a special attention has to be devoted to cultural heritage, especially churches, and that design and analysis tools and guidelines should be set up or updated.

3 Seismic Classification and Seismic Code

As said in the introductory paragraph, a large gap was developed in the last two decades between the state of art and the state of practice, or, in other words, between consolidated scientific advancements and seismic standards and zonation. This came out with great evidence when observing that S.Giuliano di Puglia was not in the current seismic zonation, while it would have been, in the 2nd seismic category, according to the 1998 proposal. Therefore buildings in S.Giuliano, as well as in most of the municipalities of the damaged areas, were designed until 2002 for gravity loads only and no seismic criteria.

The Italian Government considered the existing gap between scientific knowledge and its translation into risk mitigation tools unacceptable.

It established a Working Group to obtain scientific advices on the possible upgrade of both seismic zonation and seismic code.

The proposal was prepared in a very short time and on march 20, 2003, the Prime Minister issued, with an unusual emergency measure, the by-law n. 3274 (OPCM 3274, 2003) concerning "initial items on the general criteria for seismic zonation of the national territory and on seismic code". The ability of the working group to complete the new zonation and the new seismic code was related to the availability of document-proposals set up in the previous five years.

As far as the zonation is concerned, the 1998 map was practically adopted as a reference map for the Regions, which actually have in charge the power and the

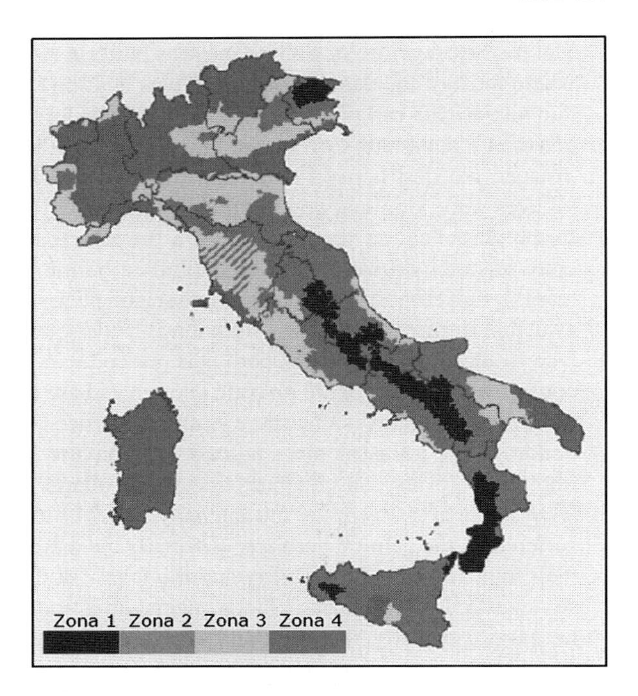

Fig. 7 Current Italian seismic zonation as determined by the by-law 3274/2003 (See also Color Plate 7 on page 211)

responsibility for the seismic classification of the regional territory. Some general criteria on the use of this reference map and on the studies to be carried out to make hazard and zonation maps in the future were also drafted and enforced with the by-law 3274. The resulting seismic zonation map after the enforcement of the Regions is reported in Fig. 7.

According to this zonation, there are 4 seismic zones, from zone 1, with the highest hazard (reference peak ground acceleration on stiff soil for 475 years. return period equal to 0.35 g), to zone 4, with the lowest hazard (reference peak ground acceleration on stiff soil for 475 years return period equal to 0.05 g), where simplified rules for seismic design can be used. Therefore, no part of Italy can be considered totally free from seismic hazard. On the other hand the areas included in zones 1–3 are almost 70% of the Italian territory, while they were about 45% in the previous classification (1984) and 25% in the classification until 1981, as said above.

As far as the seismic code is concerned, reference was made to a draft proposal for a code, based on the Eurocode 8. However the working group produced a considerable work to improve that version, taking into account the most recent release of EC8 at that time (CEN 2003).

The code was introduced at an experimental stage and its application was not compulsory until a given deadline, that was delayed many times. This, however, permitted to make considerable improvements, included in a subsequent version of the code (OPCM 3431, 2005), that ensued from the experimental applications and further studies carried out by the NCPD Competence Centres (see below).

Apart from the radical change of approach of the new norm, which is performance-based rather than prescriptive, as the previous one (Ministry of Public Works 1996), the most important and qualifying novelties, with respect to the previous Italian seismic code, can be summarized as follows:

- seismic action is described in terms of elastic response spectra and peak ground acceleration with defined probability of exceedance;
- effects of soil amplification are considered to modify both shape and intensity of the spectrum;
- expected performances of the structure are clearly stated;
- influence of structure characteristics, like geometry, regularity and constructive rules, on ductility are clarified;
- influence of ductility on the design action is clarified;
- rules to consider fragile and ductile components are given;
- rules of capacity design are adopted;
- new techniques and technologies for seismic protection, such as seismic isolation and energy dissipation, are accepted and dealt with;
- non linear analysis, alternative to linear analysis, can be used and are ruled out;
- proper attention is devoted to existing buildings, for which no a-priori assumption can be made on the overall ductility and on the performance of fragile elements; a new and original approach, with respect to EC8, is set up in the Italian code.

The NCPD, with the collaboration of its competence centres, supported the enforcement of the seismic zonation and code of by-law 3274\2003 in order to facilitate their implementation and use, by:

- replying to technical-administrative questions requiring suitable interpretation,
- revising and updating seismic rules for the future code versions,
- participating to the drafting of the Guidelines for cultural heritage,
- supporting the management of the relationships with external institutions,
- organizing refreshing courses for public administration personnel and for professionals involved in the process of seismic risk reduction.

Although the application of the new seismic code remained voluntary in the subsequent years, it is important to underline how its enforcement at an experimental level has promoted a process of total renovation of the approach to seismic design in Italy.

In recent time, new seismic hazard maps have been made available by INGV (see below), which are perfectly consistent with the criteria of the by-law 3274/2003. These maps express the seismic hazard in terms of spectral ordinates for different vibration periods and for different return periods. They have permitted to set up a new approach to the definition of the design seismic action as a function of the geographical coordinates of the site under consideration, of the lifetime of the structure and of the importance of the structure. The design seismic action thus becomes theoretically independent of the seismic classification and varies as a continuous function of the geographical coordinates, all along the Italian territory, and of the return period.

Such a new approach has been incorporated in the most recent seismic code, which is the result of a close collaboration between the Ministry of Infrastructure and the NCPD. It was enforced at the beginning of 2008 (Ministry of Infrastructure 2008). Its adoption is compulsory for public buildings and infrastructures, while it is voluntary for private buildings until June 2009.

The new seismic code, which is part of a more general technical norm for constructions, inherited all the novelties introduced with the OPCM 3274/2003 or, better, their developments consequent to their experimental applications and recent results of the research of the Centres of Competence of NPCD.

4 Guidelines for Microzonation

An important aspect, which again was emphasized by the S.Giuliano Earthquake, is related to the need of microzonation studies and maps for currently or potentially inhabited areas, in order to improve the seismic risk evaluation and to rationalize the land use, for urban and emergency planning, as well as to help designers to correctly select the seismic amplification factors.

The influence of the local soil conditions on the general consequences of an earthquake is well recognized in Italy since the 1908 Messina-Reggio Calabria earthquake (m 7.2, 86,000 fatalities). In 1909 the first Italian seismic code (Regio Decreto 18 aprile 1909, n. 193) prohibited the construction or re-construction on the "soils near or on fractures, landslides or soils prone to collapse or transmit rioting actions to the buildings, due to different geological features or different strength of the single parts of them ...".

However most of the microzonation studies carried out in Italy and their enforcement occurred only after a destructive earthquake, for reconstruction purposes in the damaged areas. This was the case of the microzonation carried in Friuli, after the 1976 Earthquake, in Irpinia and Basilicata, after the 1980 Earthquake, in Marche and Umbria, after the 1997–1998 earthquake sequence, in Basilicata, after the 1998 Earthquake. The same occurred after the 2002, Molise-Puglia earthquake. In Fig. 8 there is reported the microzonation map of S.Giuliano, set up for the reconstruction.

For each study different criteria were used, according to the state of art knowledge at the time it was carried out, but also to the scope and to the time and funds available.

Some microzonation studies were also carried out independently from the occurrence of earthquakes, in some Regions, such as Lombardia, Emilia-Romagna, Toscana, Marche. Some Regional guidelines were also drafted.

Seismic microzonation is, by law, a concern of the Region administrations. The main drawback of this ruling is that different Regions could use different criteria, so that inhomogeneous seismic safety could result for citizens living in different Regions. This is why it is important to set up nationally unified criteria, in order to harmonize approaches for urban and emergency planning.

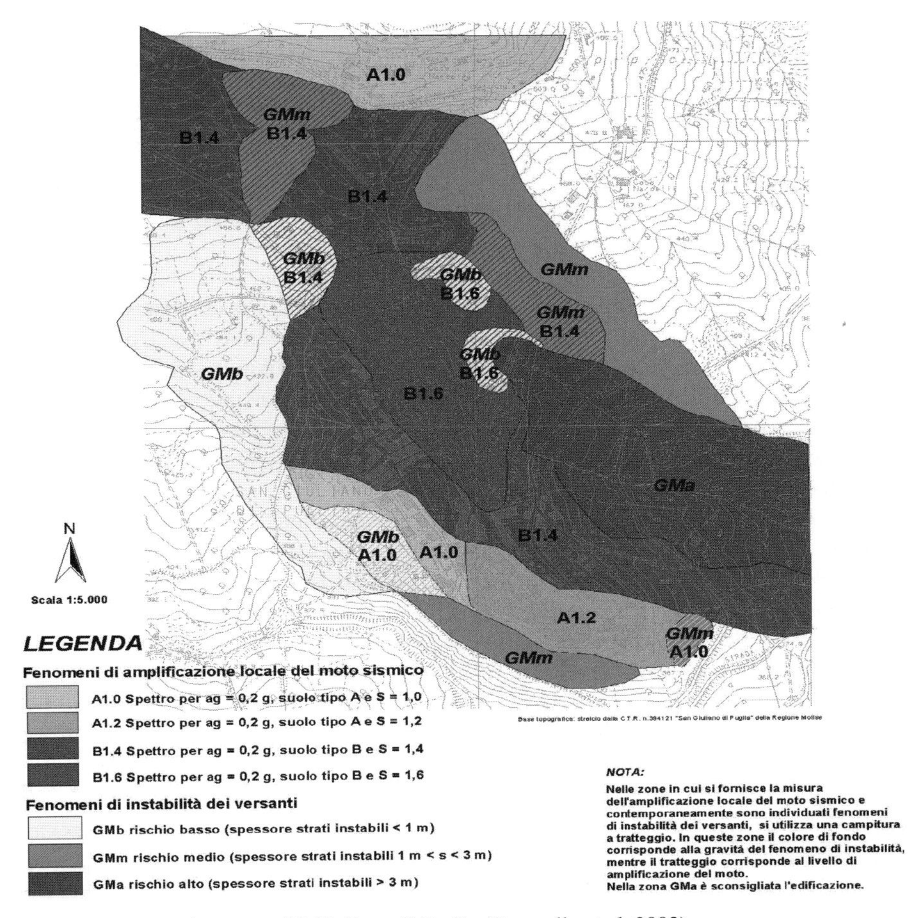

Fig. 8 Microzonation map of S.Giuliano di Puglia (Baranello et al. 2002)

The Guidelines for Seismic Microzonation (Bramerini et al. 2008) were prepared by a mixed State – Regions working group, including about 100 experts. The working group re-examined and synthesised a number of experiences carried out from 1976 until now, as well as the progresses on the subject carried in other Countries.

The Guidelines examine and quantify the role of the local site conditions in modifying the seismic shaking and causing permanent effects in the soil (liquefaction, settlements, surface faulting, landslides). This Guideline addresses land use planning at various scales, the impact of soil nature on emergency planning and on design.

The Guidelines adopt a modular approach, with reference to three levels of studies:

- level 1 → homogeneous microzones (qualitative),
- level 2 → homogeneous microzones (quantitative parameters),
- level 3 → level 2+ local in-depth investigations.

The three levels are calibrated with respect to their specific use and relevant objectives. For vast area land planning the first level is sufficient, while level 2 or level 3 are usually needed for accurate urban and emergency planning and for structural design.

5 Guidelines for Evaluation and Mitigation of Seismic Risk to Cultural Heritage

According to the actions planned by the OPCM 3274, a working group, including representatives of the Ministry of Cultural Heritage and the Civil Protection Department, as well as experts from various universities, was enforced to draft the "Guidelines for Evaluation and Mitigation of Seismic Risk to Cultural Heritage". These Guidelines are consistent with the new seismic code. They have then been enforced as a Directive of the Prime Minister (Directive of the PCM 12 October 2007).

The Guidelines have been drafted with the scope of specifying a procedure for the knowledge of the historical development of monumental buildings, for the evaluation of their seismic safety and for the design of seismic upgrading interventions.

These latter shall be consistent with the interventions on normal masonry buildings, as stated by the current seismic code, but taking into account the exigencies and peculiarities of the cultural heritage. The final judgment on the safety and the conservation of a monumental building, as guaranteed by the eventual upgrading intervention, shall be as objective and correct as possible.

The need for such Guidelines arise from the high seismic vulnerability of most of the monumental buildings and from the observation of the inadequateness of some modern interventions on Cultural Heritage, both because of their strong impact and invasiveness and of their ineffectiveness, according to the sometimes observed behaviour after recent earthquakes.

Three levels of evaluations are considered in the Guidelines, according to the level of detail of the investigation and of the numerical model, the lower level being compulsory for all the goods.

The Guidelines also devote much attention to the selection of the upgrading interventions, which should be consistent with the intrinsic characteristics of the goods, not only from an aesthetical point of view, but also with regards to the specific seismic behaviour and the safety resources of a given type of structure. These resources should be magnified, rather than neglected and replaced by different ones leading to unnatural seismic behaviours.

6 Prevention Program for Vulnerability Reduction

In order to start an effective prevention program based on the retrofit of existing constructions, some estimation of the total number of elements at risk and their seismic vulnerability and seismic risk has to be done. Such estimations can be made at different levels of detail, starting from very rough estimates. These latter can provide overall figures in order to understand the size of the problem and the way it can be tackled at national level. However, when prevention strategies must be set up, a more detailed knowledge of vulnerability and risk of single constructions is needed.

The overall seismic risk of dwelling buildings is relatively well known, however the problem is so vast that it cannot be economically sustained by the State. Incentives should be provided or insurance policies should be implemented in order to start a serious prevention program for private dwelling buildings. Obviously the awareness and the responsibility of the single citizen have to be called upon.

A different question is determined by public buildings and infrastructures. Improving the characteristics of some structures results in greater common advantage. The recognition of the seismic safety level of important structures was made compulsory for the owner with by-law 3274/2003. Also a deadline was put in five years, that has recently been delayed until the end of 2010.

A multi-step procedure was set up with OPCM 3274. The constructions that must be compulsorily checked are:

1. Buildings and infrastructural constructions of strategic importance, i.e. whose operability during and after seismic events is fundamental for the civil protection scopes;
2. Buildings and infrastructural constructions which can assume great importance in relation to the consequences of their collapse.

It is worth to mention that also the cultural heritage is considered to belong to the second category.

To get an order of magnitude of the problem, reference has been made to a census of the public buildings in Southern Italy, carried out in 1996.

Forty thousand buildings were surveyed. An extrapolation of the results to all the Italian territory was made, leading to an estimate of about 75,000 buildings to be assessed, being designed with no or inadequate seismic provisions, 35,000 of them being located in zones 1 and 2.

In order to start the verification program, a fund for interventions and seismic verifications aimed at improving the knowledge of the seismic vulnerability and reducing it, was established. 273 M€ were allocated for the years 2003, 2004 and 2005. With a part of these funds more than 7,000 seismic safety verifications and more than 200 retrofitting interventions have been funded (Dolce et al. 2008). Though significant, these figures represent only a small part of the strategic and relevant constructions potentially involved in this process.

Obviously schools deserved a special attention after the S.Giuliano Earthquake. For this reason a special program for their seismic safety improvement was enforced. About 42,000 public schools exist in Italy. 26,000 schools are located in seismic zones 1, 2 and 3. Their retrofitting costs are of the order of several billions of euros. Even concentrating the attention on the schools at greater risk in zone 1 and 2, the funds needed are of the order of four billions of euros. The initial program was funded in 2003, with an allocation of about 500 million euros resulting in about 1,600 seismic upgrading interventions.

7 Scientific Research and Civil Protection

Generally speaking, the link between Civil Protection and the scientific community has important synergic implications. On the one hand, as far as the CP viewpoint is concerned, five good reasons for a strong connection between research and CP are related to (Dolce 2008):

1. Reaching a wide scientific consensus on evaluations that imply large uncertainties, needed however to take prevention measures or to make sounded decisions to minimise the consequences of catastrophic events;
2. Making right choices for the optimisation of the resource allocation for risk mitigation;
3. Getting precise and rapid forecasting, to undertake as much fast and effective as possible warning and rescue actions;
4. Using advanced operational tools which improve the effectiveness in search and rescue operations and, more generally, in post event activities;
5. Optimising resources and actions for emergency overcoming.

Since the end of the seventies the Civil Protection has been supporting research in order to improve the knowledge on Seismology and Earthquake Engineering. Both seismological and earthquake engineering research programmes were funded, involving the whole scientific community. A strong impulse was thus given to these research areas, that would not have, otherwise, significant chances of growing up. The results of these activities can be summarised as a widespread growth of the scientific knowledge on earthquakes, both in the seismological and engineering fields, and as the progressive development of a systematic approach to the definition of seismic risk, as well as to the design of new constructions and the retrofit of the existing ones. However, not all the research products were of immediate civil protection use, as many of the sectors involved still needed a progressive development of some basic aspects.

After the S.Giuliano Earthquake It was understood that the research activities had to be re-organised and re-oriented in order to meet the actual needs for the improvement of the prevention actions.

For this reason the scientific support to NCPD was re-organised so that it is now provided by the following Competence Centres for Seismic risk:

- INGV, the National Institute for Geophysics and Volcanology;
- ReLUIS, a network of research laboratories of earthquake engineering;
- EUCENTRE, a non-profit Foundation participated by NCPD;

INGV is also the institution which has the seismic surveillance of the Italian territory in charge, implementing and managing the national NCPD-INGV seismometric network. Data are collected and elaborated in the 24-hours-active INGV operational centre (Centro Nazionale Terremoti) (Fig. 9).

Due to the importance of having an immediate and correct evaluation of earthquakes, a considerable impulse was also given by NPCD to the growth of the network and its real time data transmission and processing. At present the National

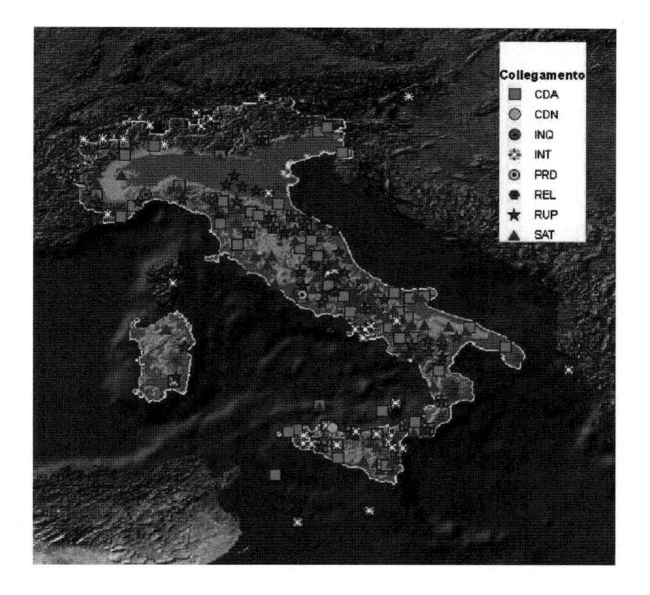

Fig. 9 Italian INGV seismometric network (See also Color Plate 8 on page 212)

seismometric network includes about 300 stations, many of them endowed with broad band seismometer, 3D strong motion accelerometer and GPS. The telematic link with the centre assures the ready and reliable availability of records, so that in a couple of minutes a first estimate of magnitude and geographical coordinates of the hypocenter is obtained and immediately reported to the operational room of NCPD.

The two Competence Centres for earthquake engineering, ReLUIS (www.reluis.it) and EUCENTRE (www.eucentre.it), have a strong attitude towards the experimental research and the training of young scientists and professionals. The experimental work has always been the weak point of the earthquake engineering research in Italy, due to the lack or inadequateness of the facilities. In the last decade, however, four university laboratories have been renewed and/or upgraded: University of Naples "Federico II", University of Basilicata – Potenza, University of Pavia and University of Trento. These laboratories have now the facilities to make dynamic (shaking table) and pseudo-dynamic (reaction wall) tests on large full-scale structural models (buildings up to 4–5 stories) as well as on large full-scale seismic devices. The ReLUIS consortium put them together in a network that, after an ad hoc agreement, also includes the Dynamic Laboratory of ENEA- Casaccia.

The EUCENTRE foundation is an international reference centre for both training, also due to the synergy with the international Rose School, and the research on specific seismic topics. EUCENTRE manages and implements the large lab facilities available at the University of Pavia, that were partly funded by NCPD.

Scientific projects of the competence centres are funded by NCPD, on a three-years contract basis. Their objectives are oriented, and relevant activities are monitored, by NCPD, in order to finalise scientific results towards products of immediate use for Civil Protection purposes. Scientific projects carried by INGV and ReLUIS involve a large number of researchers, in all the universities and institutions with high qualifications in seismology and earthquake engineering research. This is important to get results receiving high consensus in the relevant scientific community.

The research activities and the relevant results can be grouped in the two large areas of seismology (carried by INGV) and earthquake engineering (carried by ReLUIS and EUCENTRE).

7.1 Seismological Research

The 2004–2006 programme, whose activities were completed in 2007, encompassed six projects (S1 to S6; http://legacy.ingv.it/progettiSV/). The projects led either to immediately operative products or to results that are preparatory to tools for more detailed and reliable evaluations or predictions of hazard and risk. Among the immediately operative products, the results of projects S1 (Meletti et al. 2007) and S5 (Faccioli and Rovelli 2007) deserve mentioning, as they consist of the national hazard maps in terms of peak ground accelerations, pseudo-acceleration response

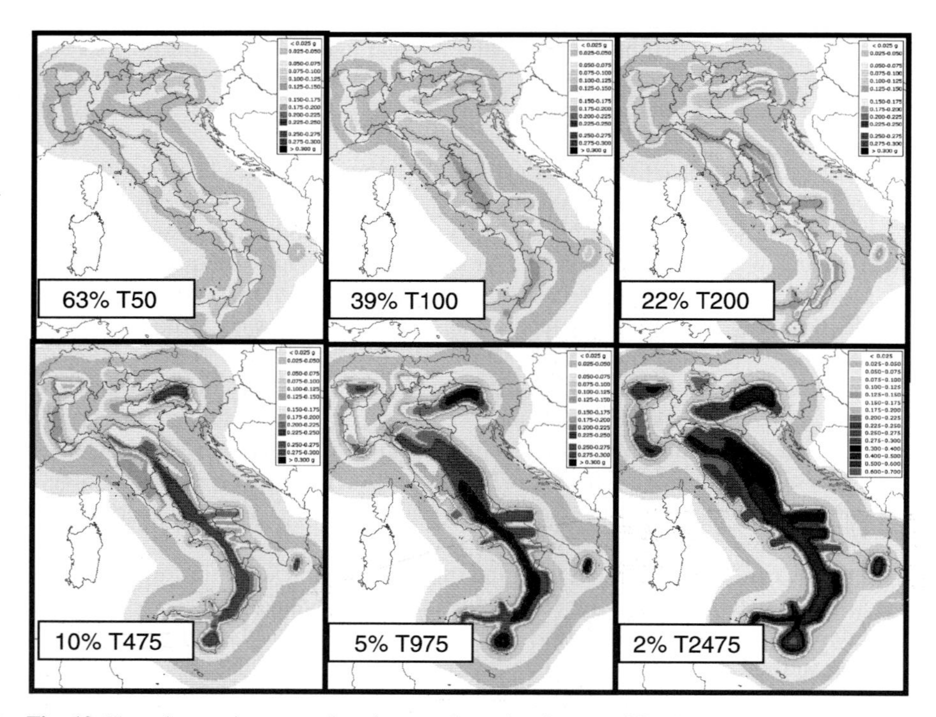

Fig. 10 Hazard maps in terms of peak ground acceleration on stiff soil (ag) for different return periods (project DPC-INGV-S1) (See also Color Plate 9 on page 212)

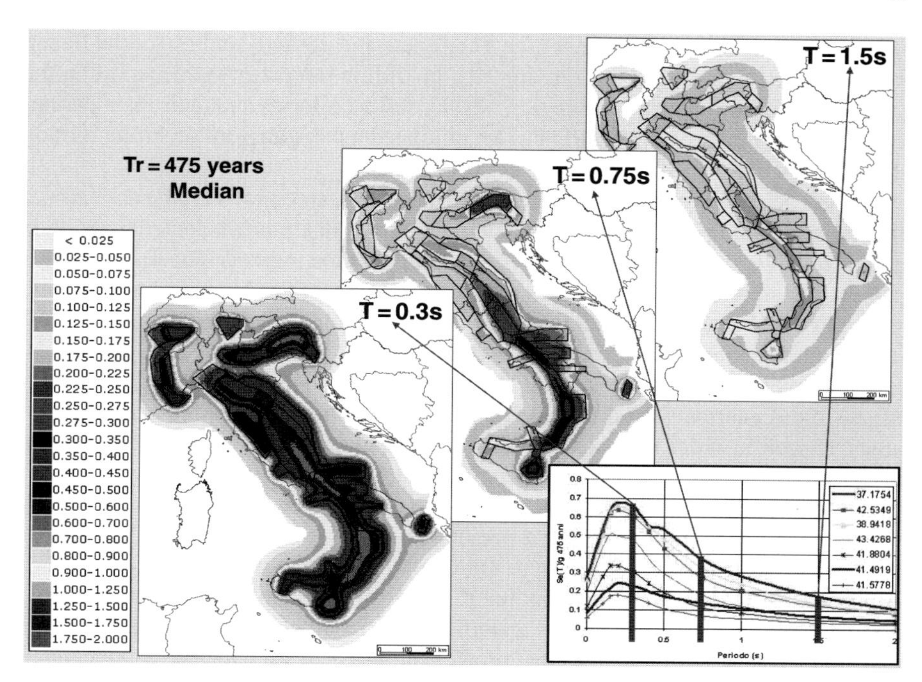

Fig. 11 Hazard maps in terms of elastic response spectra on stiff soil (project DPC-INGV-S1) (See also Color Plate 10 on page 212)

spectra on stiff soil and displacement response spectra, for different soils and return periods (from 30 to 2475 years; see Figs. 10 and 11).

The results of project S1 have been promptly exploited in the most recent version of the seismic code (Ministry of Infrastructures 2008), where the design seismic action is defined point by point in the map, for any geographical coordinates.

The displacement response spectra maps obtained by project S5 are not included in the new code, but they could easily be in the future. Indeed they are indispensable tools for the application of innovative methods of analysis and verification more consistent with the actual seismic structural behaviour, which is displacement-controlled rather than force-controlled. It is worth to remind that ReLUIS and EUCENTRE are currently developing these methods and making them available in a model code.

Among the other projects completed in 2007, project S4 set up a tool for the rapid generation of shake maps, starting from the identification of the seismic source and the magnitude of the earthquake. It also exploits the shaking parameters (PGA, PGV, PGD) derived from the strong motion networks managed by NCPD and INGV. The knowledge of a reliable shake map soon after (within some tens of minutes) the earthquake allows NCPD to make a reliable estimation of the possible damage to constructions and of the casualties, so that search and rescue resources can be optimally allocated and distributed on the territory.

Project S6 realized for the first time a complete and unified database of the Italian strong motion data (http://itaca.mi.ingv.it), providing important information on the stations and on the site characteristics, besides the waveforms and their main parameters. Actually since 1972 NCPD has accumulated a large amount of strong motion data with its national strong motion network (RAN).

Project S2 was aimed at extending some databases, among which the database of the Italian seismogenic sources, characterized by the maximum expected magnitude and, if reliable, the activation probability in a given time interval. The results of this project are aimed at improving the hazard evaluation in the long run, for the next generation of hazard maps, but they are also of immediate use at regional scale, for instance for scenario analyses.

Finally, Project S3 dealt with the preparation of guidelines for the development of seismic scenarios with different levels of detail; they were based on some test areas, chosen in order to make a methodological comparison in view of the set up of emergency planning or mitigation programmes at local scale.

The five projects set up within the new DPC-INGV research programme (http://www.ingv.it/l-ingv/progetti/), that is presently ongoing and will end in 2010, further develop some topics that either were not completed in the previous programme or still need to be improved, such as for the new project S3, which deals with the rapid estimation of the parameters and effects of strong earthquakes in Italy and in the Mediterranean area, or the new project S4, devoted to a further enhancement of the strong motion database, including the most recent records of RAN and of other networks, as well as more detailed data on the sites of accelerometric stations. The upgrading of the hazard maps obtained as a fundamental result of the previous programme is not an objective of the current programme. As a matter of fact, NCPD and INGV are moving towards an experimental dynamic model for the evaluation of the seismic hazard at national level and a further improvement of the knowledge of the seismogenetic potential in Italy, that represent, respectively, the objectives of the two new projects S1 and S2. Finally, the new project S5 is meant to support new activities, or to address some already existing, for the implementation of multidisciplinary monitoring systems of selected seismogenetic sources or areas, with the aim of enhancing the knowledge of the earthquake generation processes and occurrence rates in Italy.

7.2 Earthquake Engineering Research

A significant part of the research programmes of ReLUIS and EUCENTRE for the 2005–2008 contract was related to (i) a practical appraisal of the new code, as well as to the setting up of possibly needed modifications, and (ii) the production of manuals and codes of practice, to support engineers in this great transformation of the code. In the meanwhile, attention has been paid to innovative technologies, as well as to design and analysis methods, for future implementation of codes and for the improvement of seismic risk assessment, post-event evaluation and early warning. The research programmes dealt with 5 great themes of earthquake engineering:

- the vulnerability of existing structures and the methods for its reduction, with particular emphasis to R/C and masonry buildings and bridges;
- the innovative design criteria, which overcome the drawbacks of the current elastic calculations and verification in terms of forces and stresses rather than on displacements and deformations, or which improve the design of less common structures, such as steel and concrete-steel structures, or develop concepts and tools for seismic-geotechnical design;
- the new technologies and relevant methods for the seismic protection of structures, from seismic isolation, passive control and semi-active control to the use of fibre-composites;
- the tools for seismic risk evaluation and for emergency management, with particular attention to the databases for risk evaluations and seismic scenarios, the definition of the priorities of intervention on vulnerable buildings, the experimental damage assessment techniques, the structural monitoring and the "early warning" techniques for strategical structures and infrastructures;
- the design of peculiar structures whose seismic behaviour was less studied in the past and therefore need in-depth experimental and theoretical studies, such as, for instance, R/C precast buildings or harbour structures.

In all the research programs a great attention was devoted to the experimental laboratory activities.

In this research programme, about 130 research units were involved; distributed in all the 40 universities where research groups are active on earthquake engineering. They actually constitute a national network of earthquake engineering research. Coordination and continuous information exchanges were assured through frequent meetings and workshops, some of them extended to a professional audience.

Finally, the training activity carried out by ReLUIS and EUCENTRE deserves mentioning. It is aimed at explaining the new methods of structural design and analysis required by the new seismic code. Some tens of thousands of professionals attended the refresher courses on the new seismic code, that complied with a predefined programme drawn up together with the NCPD. The courses were organised either by the two Competence Centres or by research groups involved in the scientific activities. Moreover, the web sites of ReLUIS and EUCENTRE provide useful tools for design, such as software, databases, scientific publications and manuals.

8 Awareness and Education

The NCPD has always taken care of improving the awareness of the population and of the decision makers for seismic risk and has always been very active in making a culture of prevention and civil protection to grow up. Several volumes dedicated to the strongest earthquakes in the past, also in cooperation with external researchers, have been written, in order to refresh the memory and understand the multifaceted aspects of past earthquakes. Furthermore, leaflets containing friendly explanation of seismic risk and behaviour rules in case of earthquake have been produced and widely distributed. It has also promoted and participated to several information campaigns in Italian schools, carried out at national level by NGOs.

In the last two years a travelling exhibition called "Terremoti d'Italia" (Earthquakes of Italy) has been organized in several Italian towns. The exhibition, which covers an area of 800–1000 m^2, contains original ancient documents and films, masterpieces of famous modern artists, multimedia presentations, modern antiseismic devices and monitoring systems. The most appealing section of the exhibition is that including equipments for the simulation of earthquake effects: (a) a 4 m by 2 m bidirectional shaking table supporting 10 fibreglass building and bridge models excited by an earthquake-like motion, (b) a fully equipped room mounted on a bidirectional shaking table, capable of reproducing real strong earthquakes while hosting groups of 10–15 adults and children. Moreover; there is a special section devoted to primary and secondary school groups, developed within the DPC-INGV project, where youngsters can learn about earthquakes and make stimulating experiments. Until now, more than 50,000 people have been visited and appreciated the exhibition.

9 Conclusion

After a destructive earthquake, the attention to the seismic risk problem raises sharply, thus favoring the undertaking of important actions for seismic risk mitigation, such as improving codes and regulations, issuing new guidelines and allocating funds for vulnerability reduction of inadequate constructions.

The 2002 S.Giuliano Earthquake was the event that pushed the Italian government to undertake a series of risk mitigation actions, whose long term influence is still active.

The seismic code and the seismic zonation were revised soon after the earthquake. This produced initially disconcert, due to the difficulties in the application of the new rules, but also an extraordinary effort at all levels to disseminate new design philosophies: from professionals to scientific and administrative structures.

An important novelty with respect to previous similar situations after past earthquakes, is that direct risk mitigation countermeasures were undertaken at national levels, by making the safety verification of strategic and relevant structures compulsory in a limited time framework and by allocating funds, though limited, to support the seismic rehabilitation of existing important constructions.

The results obtained from the safety verifications will certainly result in the enhancement of the awareness of public administrators and decision makers about the seismic deficit that certainly characterize most of the existing constructions in seismic areas and the need to increase the attention to the problem and devote larger funds to implement effective seismic mitigation policies. As a matter of fact, in order to become fully effective, actions for a long term perspective are necessary in the future, with the certainty of the availability of adequate resources.

Acknowledgments The author is indebted with all the personnel of the Seismic Risk Office of the Italian Civil Protection Department that operated in the recent years to make the activities described in this paper and produce some of the data and elaborations presented in this paper.

References

Baranello S, Bernabini M, Dolce M, Pappone G, Rosskopf C, Sanò T, Cara PL, De Nardis R, Di Pasquale G, Goretti A, Gorini A, Lembo P, Marcucci S, Marsan P, Martini MG, Naso G (2002) A Criterion for the Seismic Microzonation in Molise after the 31 October 2002 Earthquake (in Italian). Proceedings of 11° Convegno L'Ingegneria Sismica in Italia, Genova

Bramerini F, Di Pasquale G, Naso G, Severino M (editors) et al. (2008) Guidelines for Seismic Microzonation (in Italian). Presidenza del Consiglio dei Ministri – Dipartimento della Protezione Civile e Conferenza delle Regioni, Roma

CEN (2003) PrEN-1998-1 – Eurocode 8: Design Provisions for Earthquake Resistance of Structures, Part 1.1: General rules, seismic actions and rules for buildings

De Marco R, Martini MG (2001) La classificazione e la normativa sismica italiana dal 1909 al 1984, Presidenza del Consiglio dei Ministri-Servizio Sismico Nazionale, Ist. Poligrafico e Zecca dello Stato, Roma

Directive of the PCM 12 October 2007 (2007) Evaluation and the Reduction of the Seismic Risk of the Cultural Heritage with Reference to the Technical Norm for Constructions (in Italian). G.U.R.I. n. 24 of 29 January, 2008

Dolce M (2008) Civil Protection vs. Earthquake Engineering and Seismological Research. Proceedings of 14th WCEE, October 2008, Beijing, China, Keynote speech

Dolce M, De Sortis A, Di Pasquale G, Goretti A, Paoli G, Papa F, Papa S, Pizza AG., Sergio S, Severino M (2008) The Evaluation and the Reduction of Seismic Risk: National Initiatives (in Italian), 6th VGR, Pisa

Faccioli E, Rovelli A (2007) Project S5 – Seismic Input in terms of Expected Spectral Displacements – Final report (in Italian). INGV. http://progettos5.stru.polimi.it

Gavarini C (coordinator) et al. (1999) Proposal of Seismic Re-classification of the National Territory (in Italian). Ingegneria Sismica n.1

Goretti A, Dolce M (2004) Post-Earthquake Site Effect Evaluation From Damage and Building Type Data: An Overview of Italian Recent Applications. Proceedings of 13th WCEE, Vancouver, Canada

Jiménez MJ, Giardini D, Grünthal G (2003) The ESC-SESAME Unified Hazard Model for the European-Mediterranean Region. EMSC/CSEM Newsletter, 19, 2–4

Meletti C, Calvi GM, Stucchi M (2007) Project S1 – Pursuing the Assistance to DPC to Complete and Manage the Hazard Map of OPCM 3274/2003 and Design of further Development – Final Report (in Italian). INGV. Interactive Maps of Seismic Hazard (WebGis) available on http//esse1.mi.ingv/

Ministry of Public Works (1996) Decree 16 January 1996, Technical Norms for Constructions in Seismic Areas (in Italian). G.U.R.I. n. 29 of 5-2-1996

Ministry of Infrastructures (2008) Decree 14 January 2008, Approval of the New Technical Norms for Constructions (in Italian). G.U.R.I. n. 29 of 4-2-2008 – S.O. n.30

OPCM 3274 20 March 2003 (2003) Initial Items on the General Criteria for the Seismic Zonation of the National Territory and the Seismic Code (in Italian). G.U.R.I. n. 72 del 08/05/2003

OPCM 3431 3 May 2005 (2005) Further modifications and integrations of the OPCM of 20 March 2003 (in Italian). G.U.R.I. n. 107 of 10 May 2005- S.O. n.85

New Challenges in Geotechnique for Ground Hazards Due to Intensely Strong Earthquake Shaking

Kenji Ishihara

Abstract During the 2007.7.16 earthquake that rocked Kashiwazaki area southwest of Niigata city, multiple series of strong motions were recorded in excess of 1000 gals in terms of peak acceleration. The premise of the nuclear power stations nearest the epicenter suffered distortions, settlements and uneven sags on the ground surface, although no vital damage was incurred to the buildings and facilities. In the city area of Kashiwazaki, the sea walls along the coast for protecting high tides and waves were displaced largely although these seemed to be fairly sturdy. Looking over the damage features manifested, one can realize some new problems of geotechnical importance for which efforts should be expended particularly for the ground conditions and soil structure interaction subjected to intense shaking. These include the significant amount of settlement of the ground consisting of partly saturated silty sand and the likelihood of remarkable increase in earth pressure due to ratcheting type of interaction between wall structures and adjacent soil deposits. These issues will be addressed as a novel area of research in the earthquake geotechnical engineering in this paper.

Keywords Ground damage · Settlements · Earth pressure

1 Introduction

Over the last 30 years, there have been remarkable progresses in sophistication of instruments and strengthening of observation network for recording strong motions during earthquakes. In Japan there is a network of stations called K-Net consisting of about 1000 high-precision recorders placed on the ground surface with a spacing of 25 km × 25 km throughout the country. Another set of recording stations called Kick-net has also been established in which about 700 sites are installed with instruments both on the ground surface and on the bed rocks. Thus, the coverage of areas was

K. Ishihara (✉)
Research and Development Initiative, Chuo University, 1-5-7 Kameido, Koto-ku, Tokyo, 136-8577 Japan
e-mail: ke-ishi@po.iijnet.or.jp

A.T. Tankut (ed.), *Earthquakes and Tsunamis,* Geotechnical, Geological, and Earthquake Engineering 11, DOI 10.1007/978-90-481-2399-5_7,

Table 1 Increasing trend of intensity of monitored horizontal ground shaking in Japan

Earthquake	Recorded site	Year, month	Max. acceleration (gal.)
Iwate-Miyagi	Ichinoseki	2008, June	4022
Niigata-Chuetsu (aftershock)	Kawaguchi	2004, Oct.	2515
Niigata-Chuetsu	Toka-cho	2004, Oct.	1750
Miyagiken-North	Naruse	2003, July	2037
Miyagiken-Oki	Sumiyoshi	2003, May	1305
Kobe	JMA	1995, Jan.	891
Hokkaido-Nansei	Kayabe Bridge	1993, July	640
Nihonkai-Chubu	Akita	1978, June	400
Tokachi-Oki	Hachinohe	1968, May	225

widened and precision improved greatly to seize motions throughout the country during any scale of earthquakes.

As a consequence, the magnitude of recorded accelerations has increased remarkably year after year, as indicated in Table 1. At the time of the Tokachi-Oki earthquake in 1968, the peak horizontal ground acceleration recorded was 225 gal in the port of Hachinohe north of Japan, but it jumped up to a value of 891 gal at the time of the 1995 Kobe earthquake. Since then, the networks of recording stations were strengthened and at the time of the most recent earthquake in 2008, the peak recorded acceleration reached a value as high as 4022 gal in combined 3D absolute peak acceleration. Since earthquakes are natural phenomena, it seems unlikely that the intensity of motion itself has in fact increased so dramatically, but it is not permissible to ignore such strong motions once they are recorded.

From the engineering point of view, the increase in the recorded motions is to be recognized as dictating that the design for structures and facilities are correspondingly made to cope with such an increased level of seismicity. Thus, engineers are confronted with new challenges as to how to cope with such an increases demand and to come up with effective and still economically feasible concepts and procedures.

Shown in Table 2 are the evolutions with time regarding capacity evaluation in terms of concepts and method of analysis in response to the increasing demand. In the design and practice of nuclear power facilities, the accelerations in input motions had been set at a high value from the early period and equivalent linear analysis procedures have been used. The major efforts being undertaken in Japan for the nuclear facilities after the 2007 October earthquake is further strengthening of the ground surrounding foundations and water-intake facilities.

In the field of structural engineering, the psudo-static method was employed until around 1970, but linear analysis procedure using computer codes has become a common practice. This was followed by the non-linear analysis from early 1980. On the other hand, theory and practice was developed for the structural control and vibration isolation from around 1980s. Although this technique is an outgrowth from general demand to reduce the damage, it is also viewed as being helpful to cope with the very strong shaking during earthquakes.

In the area of geotechnical engineering, psudo-static analysis has long been used to ensure the safety of embankments and dams against an external force due to seismic shaking. Subsequently equivalent linear analysis procedure has been used to assess the amplification characteristics of local soil deposits and to evaluate the

Table 2 Evolution in demand versus capacity in the seismic design

Year	1970	75	80	85	90	95	00	05	10	15
Seismicity	200-300 gal		400-500 gal				600-800 gal		1500 gal	
Nuclear power facilities				Equivalent Liner analysis					Streng-thening	
Structures (Buildings, Bridges)	Psudo-static analysis	Linear analysis		Non-linear analysis						
						Vibration control Isolation				
Geotechnics (Liquefaction, Landslide, Soil structure interaction)	Psudo-static analysis (Fs≧1.2)			·Equivalent linear analysis ·Effective stress-based analysis		·Permanent displacement analysis ·Performance-based design ·Hazard map. Risk map				

deformations developed in earth structures. With the evolution of the constitutive laws characterizing soil deformations, the response analysis procedures based on effective stress principle has been developed and put into practical use. This method consists in evaluating gradually decreasing effective confining stress and reflecting it on the decrease in stiffness and strength as seismic excitation proceeds with time. Thus, this method is considered to reflect actual situations in which large acceleration is suppressed by softening of soils, but accompanied in turn with large deformations. The development of the response analysis based on the effective stress principle has exerted a great impact on the advances of the design method based on large and residual deformations. With this tool, what is called the performance-based design was vastly enhanced, as indicated in Table 2. The large acceleration corresponding to a large shear stress in excess of strength will make soils deformed largely and the design criteria for specifying an allowable deformation will become a major yardstick for the design of soil deposits and earth structures.

Another countermeasure to reduce the risk of damage due to soil failure would be to prepare local zoning maps as indicated in Table 2 and to enhance preparedness amongst residents for possible distress caused by soil failure. Efforts are being made in this direction to reduce risk of damage by large earthquakes.

Consequences of strong shaking during earthquakes to instability of the ground have been recognized widely as liquefaction of saturated sands and ensuing flow failure and settlements. However, there have been other types of damage in recent earthquakes which have not been properly addressed. The following is an introduction of new features of the ground damage observed in the recent earthquake in Japan.

The occurrence of significant settlements of the ground composed of well-compacted partly saturated silty sand fills has never been identified and documented in the literature. In the premise of the nuclear power station in Kashiwazaki, the settlement of the order of 30–50 cm did actually take place, at the time of the Niigata Chuetsu-Oki Earthquake of 2007. 7. 16, accompanied by local distortions or offsets on the ground surface. This appears to have accrued as a result of extraordinarily strong shaking of the order of 1000 gals in acceleration, inducing a peak cyclic stress ratio of 0.8–0.9 in dynamic loading. On the other hands, some laboratory triaxial tests had been conducted, at the Tokyo University of Science, on compacted samples of a silty sand to investigate the volume change and residual strength characteristics under irregular loading conditions (Sawada et al., 2006). To shed some light on what was observed at the nuclear power plant site, the outcome of the above tests was cited in this paper and a preliminary interpretation was given to the occurrence of the ground settlements.

The depression developed in the vicinity of vertical walls of the buildings in the nuclear power station suggested that there might had been gradual clogging of the openings by surrounding soils during the shaking, which must have conduced to an increase in earth pressure on the wall. This phenomenon will be cited as ratcheting action.

There was another place in the area of equally strong shaking where the sea walls were damaged due to the increased horizontal thrust against the wall which might have resulted from the ratcheting phenomenon. On the other hands, some studies had been under progress, using the model box, at the Tokyo University of Science, by Tatsuoka et al. (2008) on the increase in earth pressure by repetitive forward and backward wall movements. It was found that the ratcheting soil movement around the wall could lead to a greatly increased earth pressure which would not be able to be explained by the conventional concept. Since this effect has not been addressed properly in the past, some preliminary interpretation will be given to what happened during the 2007 earthquake in the light of the observed results of the model tests in the laboratory.

2 General Features of the 2007 Earthquake

The coastal area in the middle of Niigata Prefecture in Japan was shaken at 10:13 a.m. on July 16, 2007 by a fairly large earthquake with a magnitude of 6.8. The epicenter of the quake is shown in Fig. 1. The severely shaken area of Kashiwazaki is shown in Fig. 2 where the damage to the ground and houses was notable. In the flat area of the city 10 km long along the coast and 5 km wide to the inland, houses, buildings and several facilities were damaged severely by the strong shaking and also by liquefaction of sandy deposits.

2.1 Characteristic Features of Ground Damage at the Site of Nuclear Power Station

There were seven complexes of the Nuclear Power Station in Kashiwazaki-Kariwa, each with a capacity of 1.1 million kW electric power generation. These power stations are located in the area of sand dune 2 km long and 0.5 km wide along the

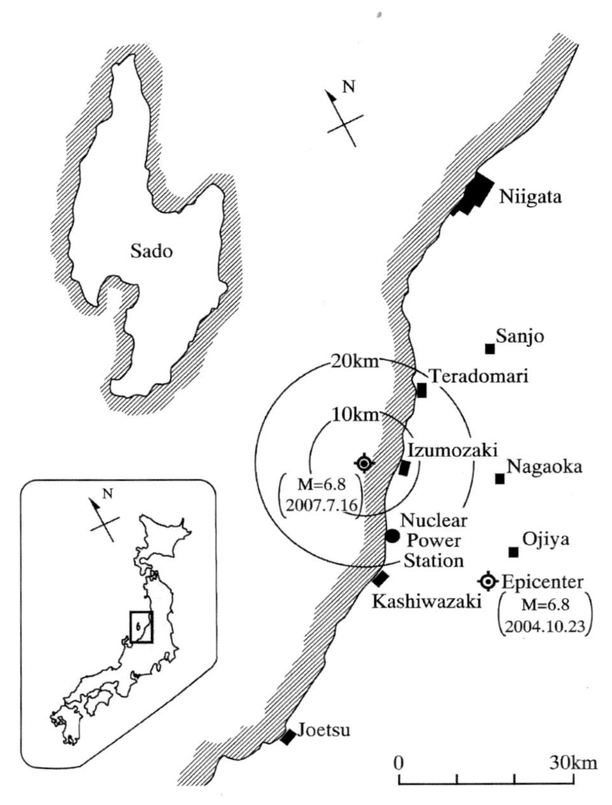

Fig. 1 Epicenter of the earthquake

coast of the Japan Sea as shown in Fig. 2 (Sakai et al. 2008). The general feature of subsurface conditions and the reactor buildings is demonstrated in Fig. 3 where it can be seen that the reactor buildings are all sitting on the base rock. More detailed arrangements of the buildings and structures within the premise of the power generating stations are shown in Fig. 4. In constructing sets of buildings, the dense deposit of sand was removed to expose the soft rock formation called Nishiyama layer, and the buildings for the reactors, generator turbines and other facilities were placed directly on top of the base rock. Then, the open space between the buildings and excavated face were back-filled with sandy soils and compacted to a sufficient density. At the floor of each reactor buildings, strong-motion seismographs had been installed and these were triggered at the time of the 2007 earthquake, bringing about seven sets of recorded motion during the quake.

The locations of the seismographs in plan are shown in Fig. 5 for the power generation station unit No.1 to No.4 with the symbol 1-R2, 2-R2, 3-R2 and 4-R2. For each of the cross sections 1–1′, 2–2′, and 3–3′ indicated in Fig. 5, the side view is shown in Fig. 6 where the location of the seismograph is also indicated. Time histories of the motions recorded at the basement of reactor building No. 1 are displayed in Fig. 7, together with the acceleration spectra of each motion 7d. It is known in Fig. 7 that the peak recorded acceleration at the 5th basement of the No.1

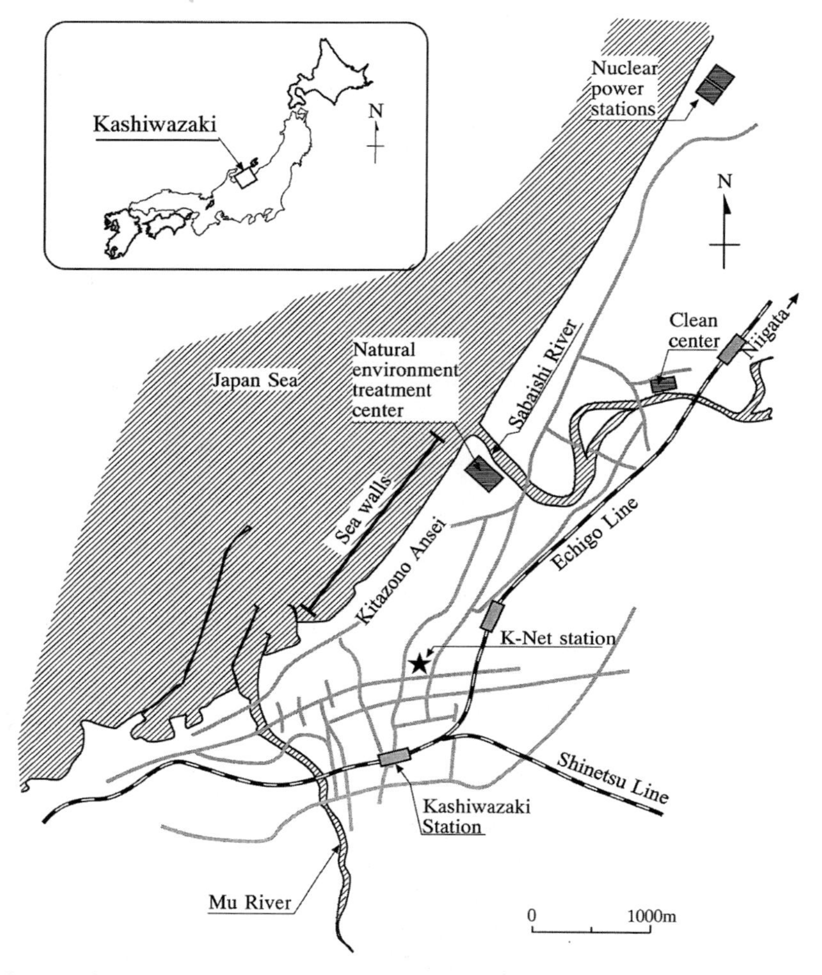

Fig. 2 Area of strong shaking and damage in Kashiwazaki city

reactor building (1-R2 in Fig. 6) was 680 gals in E–W direction and 311 gals in N–S direction. It is also noted in Fig. 7d that the predominant period in the spectrum was 0.7 and 0.8 s, in E–W and N–S component, respectively.

There were two types of ground damage of essential importance which seem to pose new problems to be explored in the realm of earthquake geotechnical engineering. These are the settlements of unsaturated compacted fills in the level ground and the depressions of the ground developed in proximity to the building.

2.1.1 Settlements of the Ground with Unsaturated Compacted Silty Sands

In the area of level ground, the settlements of the order of 10–50 cm took place in the compacted deposit of silty sand. Figure 8 shows the overall settlement of

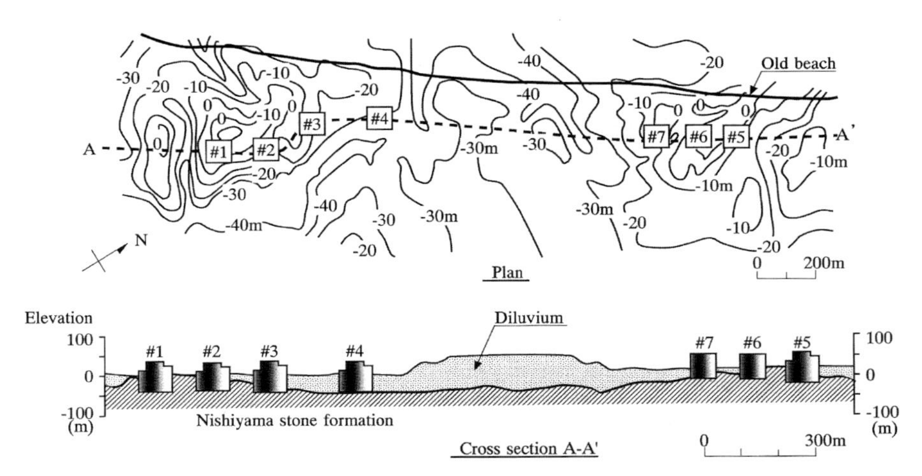

Fig. 3 Subsurface condition at the site of the nuclear power station (from Tokyo Electric Power Co.)

the yard in the premise. The ground water table had been lowered by pumping out the water and held at a level of the embedded basement of the buildings. In fact, following the earthquake, the level of the ground water was monitored at various spots, as indicated by black circles in Fig. 6, in the vicinity of the reactor and turbine buildings. The water pumping had been underway at the bottom

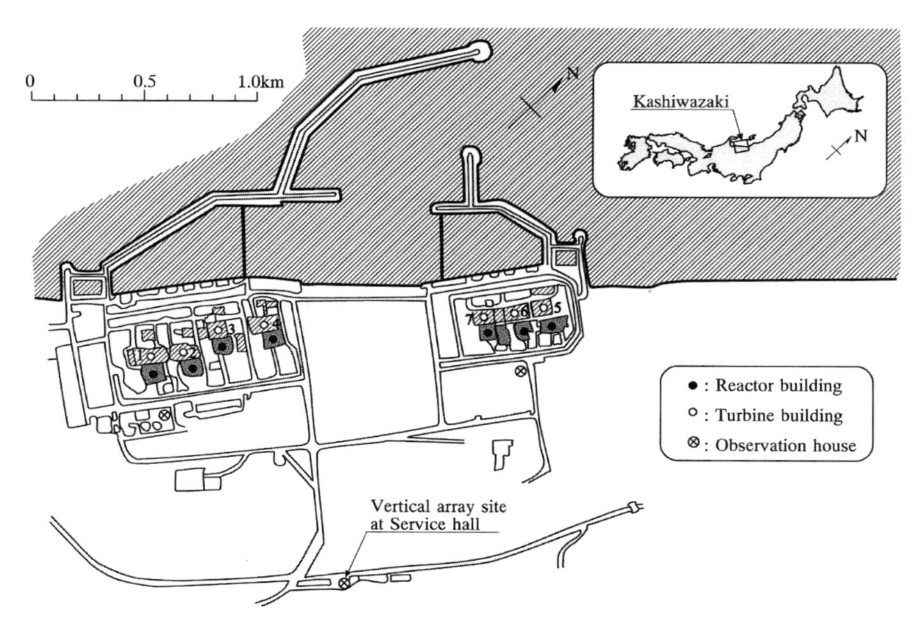

Fig. 4 Locations of the buildings at the nuclear power plant site (from Tokyo Electric Power Co.)

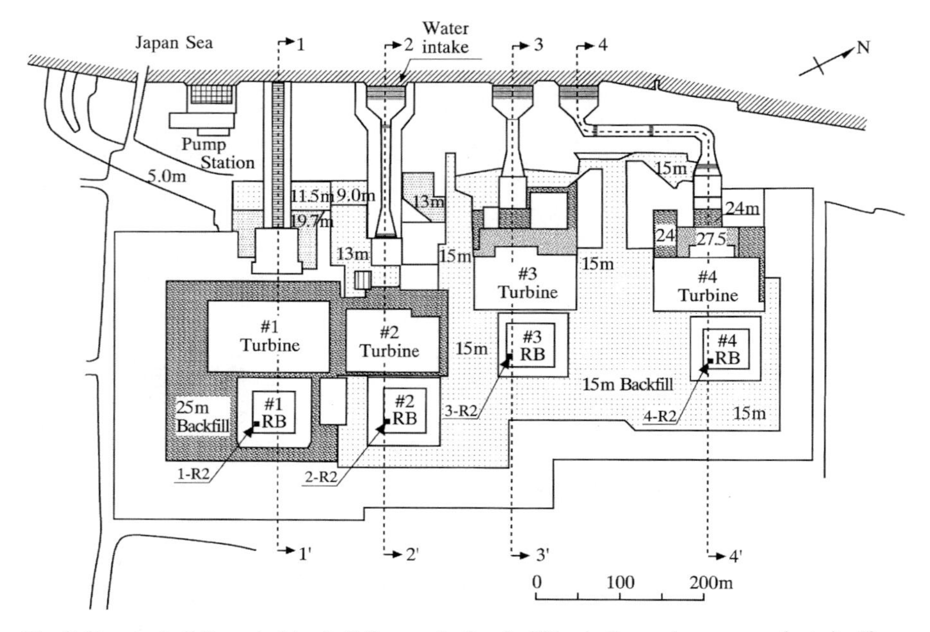

Fig. 5 Reactor buildings, turbine buildings and other facilities in the nuclear power plant site (from Tokyo Electric Power Co.)

ends of the building basement as indicated by open circles in Fig. 6. Thus, it is with reasons to assume that the silty sand in the fills was not saturated. In fact, the post-earthquake measurements of saturation ratio for the soils in the premise indicated values of $S_r = 60–90\%$. In addition, the soils were compacted fairly

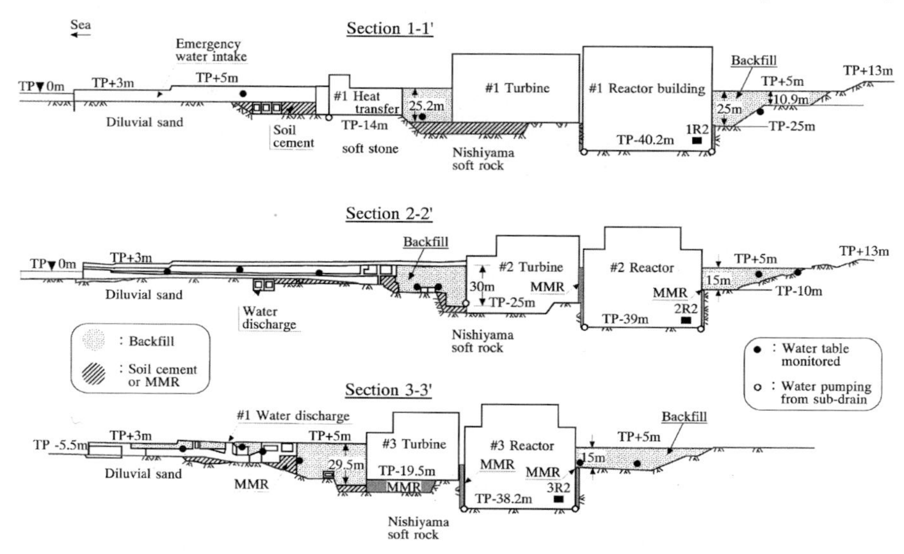

Fig. 6 Side view of the back-filled soils in the vicinity of the nuclear power facilities (from Tokyo Electric Power Co.)

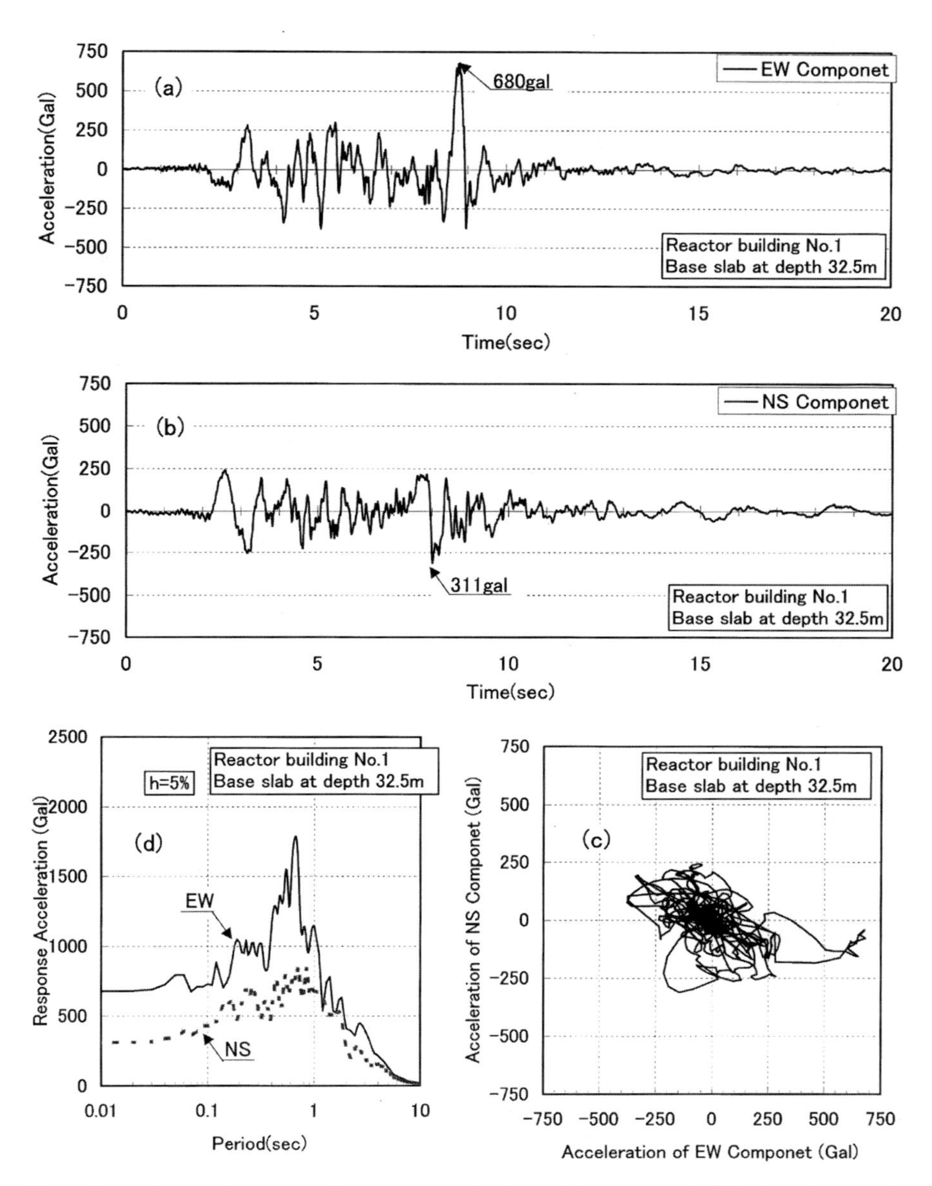

Fig. 7 Recorded accelerations and response accelerations at the basement of No. 1 nuclear reactor building (Prepared from the records of Tokyo Electric Power Co.)

dense to approximately 100% of dry density in the standard proctor energy. With all the above observations taken together, it can be conclusively mentioned that the settlements of the level ground took place in partly saturated back-filled soil deposits by the application of intense shaking on the order of 500–1000 gal on the ground surface.

Fig. 8 Settlements of the ground at the site of nuclear power station

2.1.2 Ratcheting Depression and Earth Pressure Increase

When a soil deposit in contact with a vertical rigid wall is shaken horizontally, the wall will be separated instantaneously from the surrounding soil deposit as schematically illustrated in Fig. 9. Then, the soil will fall down into the aperture and fill the gap. At the next instant, the in-filled soil will be compressed and packed by the relative movement of the outer soil mass towards the wall, thereby exerting a greater earth pressure on the wall. When this process is repeated many times, the amount of soil filling the gap tends to increase, and would eventually produce a state of an earth pressure which is by far greater than the pressure at rest, or even more than the pressure normally considered in the design of walls against seismic shaking.

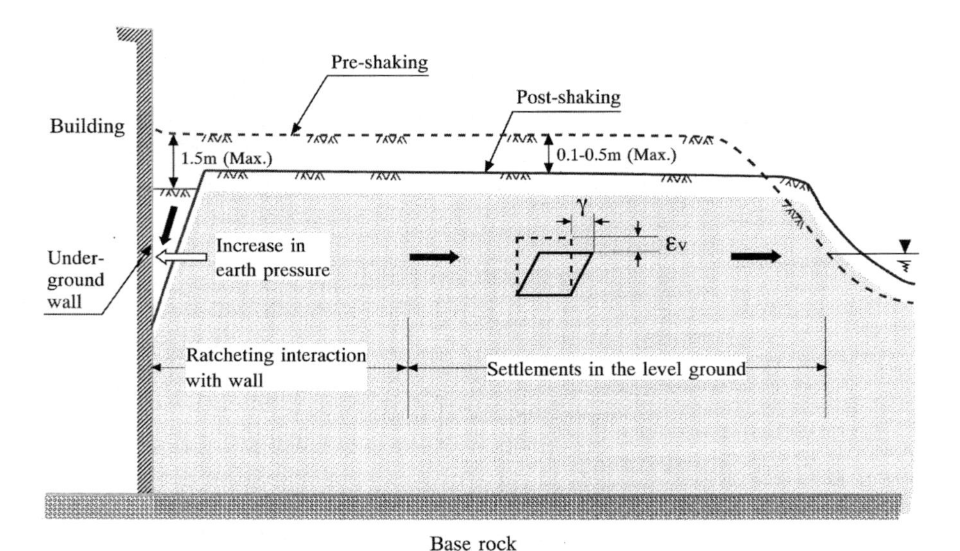

Fig. 9 Settlements and ratcheting soil-wall interaction at the site of nuclear power station

Fig. 10 Depression of the ground in the vicinity of the building

It is also to be noted that there remains a depression in the soil deposit in close proximity to the wall. Figure 10 shows the settlement near the building. If there are cables or pipes entering the building at shallow depths, these lifelines will be bent or pulled apart and exposed to danger of breakage. This kind of phenomenon called soil-structure ratcheting was observed at many places near the buildings in the premise at the time of the earthquake.

2.2 Damage to Seawalls Along the Coast of Kashiwazaki

The near-coast area, called Ansei and Kitazono, north of Kashiwazaki city is a flat low-lying zone enclosed by natural levees of two rivers and the coastal dune. In order to provide protection against high tides and ocean waves, the coastal dike had been constructed to a level of about 7 m above the sea. The location of the coastal dike is shown in Fig. 2 and its enlarged map is shown in Fig. 11. The whole body of the soil in the dike was retained, on the seaside, by the reinforced concrete parapet walls about 50 cm thick at its top. There were stacks of water-breaking blocks placed

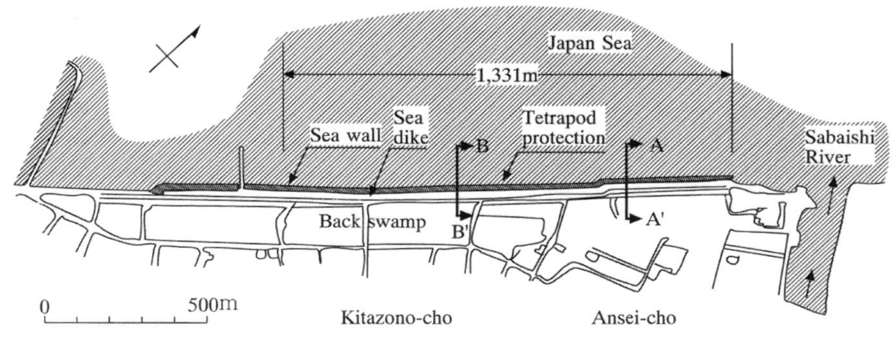

Fig. 11 Sea walls at Ansei-Kitazono area

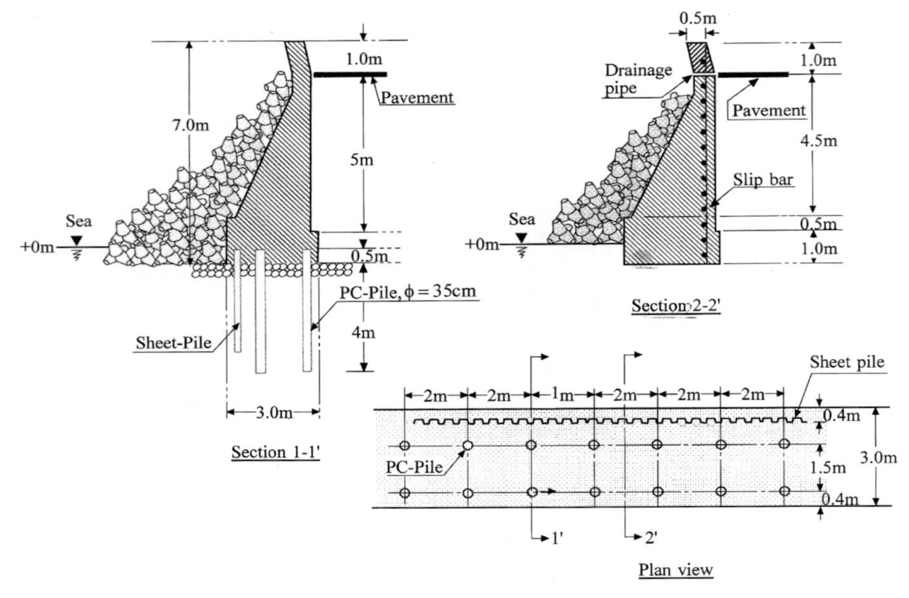

Fig. 12 Section A–A′ for the parapet wall at Ansei-cho (Nagaoka District Development Bureau, 2008)

in front of the parapet walls to provide counterweight and also for wave breaking purpose. The landside of the dike is a swampy area where there are rice-fields and small scale factories now in use. The cross sections and plan view as the walls were designed are shown in Fig. 12 for Ansei-cho section. It may be seen that the buttress type wall 7 m high were supported by rows of PC-piles 35 cm in diameter and 4 m long. The wall was counterbalanced by stacks of concrete blocks (Fig. 13).

By the strong shaking during the 2007 event, considerable damage was incurred to the tide-protecting reinforced concrete wall and the coastal dikes behind it. The features of the destruction are demonstrated in two cross sections shown in Fig. 14a, b where it can be seen that the rigid parapet seawall counterweighted by the concrete blocks was displaced by 1.0–1.5 m towards the sea. The height of the wall was 5.8–7.0 m and weight of each wave-braking block was 4 or 5 tons. The length through which the parapet wall was constructed was 1330 m as indicated in Fig. 11. Figure 15 shows the damage to the wall and the wave-braking blocks. There was an opening about 0.5–1.0 m wide developed just behind the parapet wall as displayed in Fig. 16. The pavement behind the wall also experienced a settlement of about 30 cm. In the light of the fact that the base of the wall was protected with heavy counterweight by the robust wave-braking blocks, the outward movement of the base is considered to have been small as compared to the wall displacement at the top. Thus, the parapet wall appears to have tilted towards the sea as illustrated in Fig. 14. At some locations, the pavement was covered by the liquefied sand ejected from the opening as shown in Fig. 16. On the other hands, the counterweight has

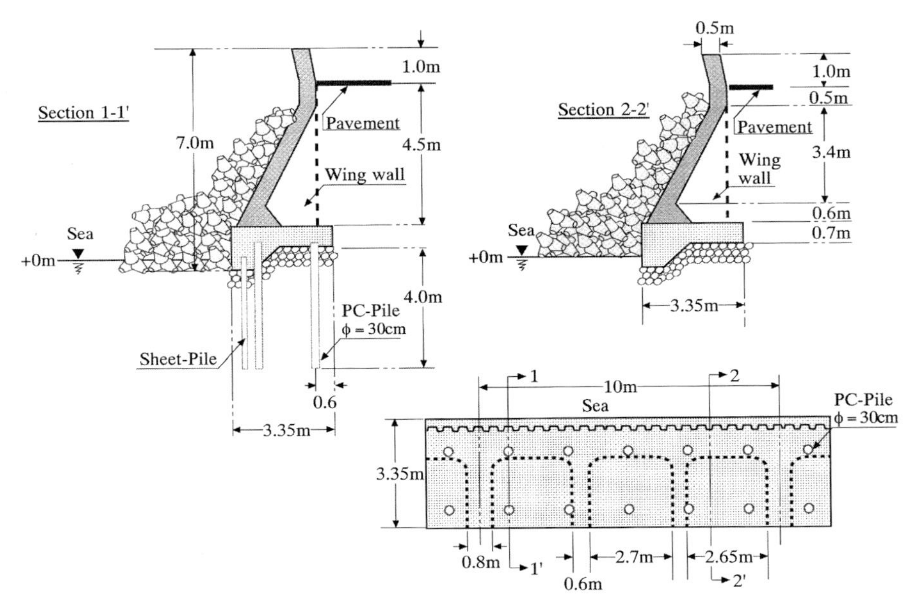

Fig. 13 Section B–B′ for the buttress type parapet wall at Kitazono (Nagaoka District Development Bureau, 2008)

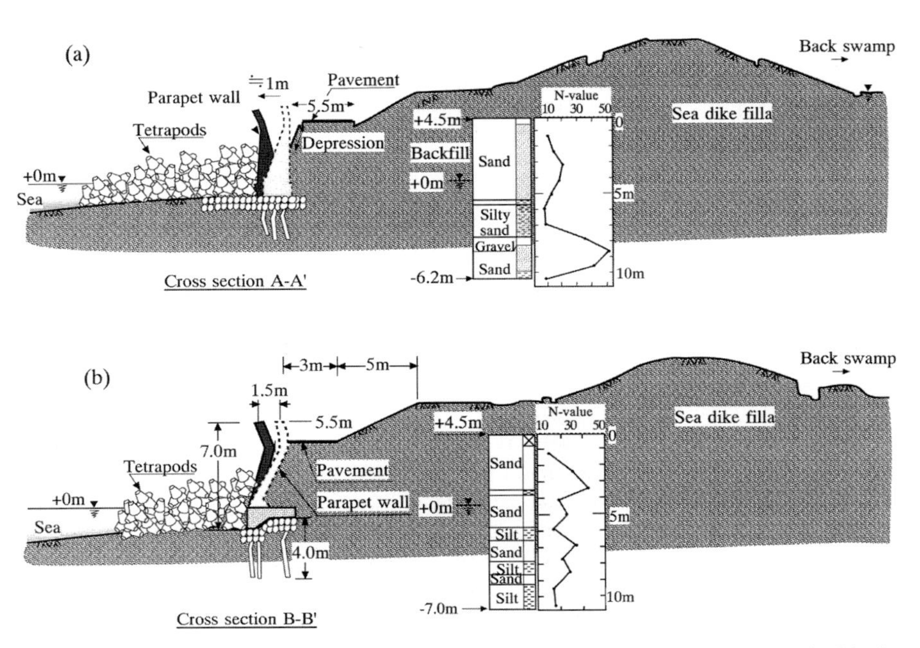

Fig. 14 Cross section of the seawall and sea dike at Ansei-cho and Kitazono-cho (Nagaoka District Development Bureau, 2008)

Fig. 15 Stacks of wave-braking blocks and parapet wall displaced about 1.0–1.5 m seaward at Kitazono area

been known to be very effective in protecting the seawalls by the seismic force from the experience in the Kobe earthquake in 1995. Thus, it was surprising to observe that the similar walls had suffered the substantial damage involving the seaward tilting with the movement as much as 1.0–1.5 m. Causes of the damage in the seawalls as above may be envisaged as being two-fold as described below.

(1) As a result of detailed soil investigations by Nagaoka District Development Bureau (2008), liquefaction was cited as one of the causes leading to the damage to the parapet walls. In fact, as seen in the soil profiles shown in Fig. 14a, b, the silty sand or sand layer with a SPT N-value of about 10 was found to exist generally at the depth 1–4 m below the sea level in the area affected. Thus, an overall movement of a large soil mass is considered to have occurred along the liquefied basal sand layer as illustrated in a mosaic pattern of displacement in Fig. 17. The overall movement of the soils above the liquefied zone is considered to have induced an intense earth pressure resulting in the seaward movement of the walls.

Fig. 16 Opening about 0.5–1.0 m wide behind the sea wall and sand ejected as a result of liquefaction in Ansei area

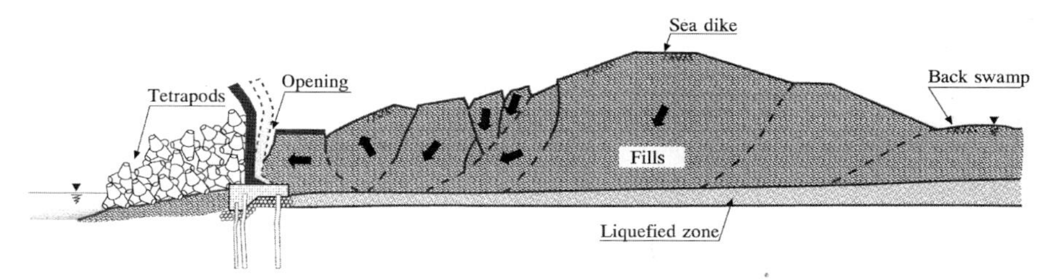

Fig. 17 Mosaic pattern of soil movement in the sea dike fill (Nagaoka District Development Bureau, 2008)

(2) As mentioned above, there was a trace of intensive interaction between the body of the seawall and backfilled soils as evidenced by openings and settlements of the pavement apron just behind the wall. In addition, mode of movement of the parapet wall was such that its top was tilted towards the sea. From these observations, it appears likely that the ratcheting action in which a fissure is opened and closed with successive clogging by soils behind must have taken place during the shaking. Thus, with the consequent increase in earth pressure, the sea wall is envisaged to have titled towards the sea.

2.3 Volume Change of Unsaturated Silty Sand in Seismic Loading

To investigate the volumetric strain characteristics of partly saturated soils subjected to seismic loading, a series of the laboratory tests had been performed at the Tokyo University of Science. The results of the tests using a triaxial test apparatus were reported by Sawada et al. (2006). The material tested was silty sand secured from the alluvial deposit in Ohgishima, Kawasaki, Japan. It contains fines of $Fc = 22\%$ and D_{50} is 0.213 mm. The maximum dry density by the Standard Proctor Compaction test was $\rho_d = 1.73\,g/cm^3$. This material was compacted in the triaxial cell to a void ratio of about 0.75 and 0.63 with the saturation ratio of Sr $\cong$ 50 and 75%. Then, the samples were first consolidated isotropically to a confining pressure of $\sigma_o' = 98\,kPa$ and then an additional static axial stress of $\sigma'_s = 42\,kPa$ was applied to produce a state of anisotropic consolidation with $K_c = 0.7$. The triaxial test apparatus used was equipped with an inner cell having a narrow mouth to permit monitoring of volume change during and after the dynamic load application.

The dynamic axial load was then applied undrained to partly saturated samples with an irregular wave form as shown in Fig. 18a. This time history was the record obtained at the time of the 2001, March 24 Geiyo Earthquake which rocked Yamaguchi and Hiroshima region in Japan. In one of the test series, the irregular axial load was applied to the specimen so that its peak, $\sigma_{d\,\max}$, could be realized in the triaxial compression side (CM-test). In another series, the direction of the irregular time history was applied reversely so that the peak could be directed towards the triaxial extension side (EM-test). The results of the CM- and EM-tests were taken

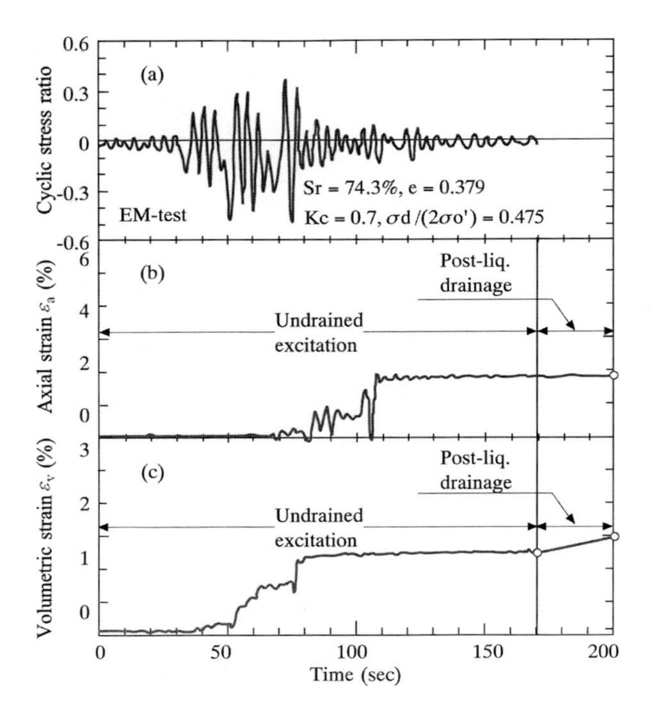

Fig. 18 Typical results of dynamic loading triaxial test with an irregular time history of axial load

simply as being from different tests with different wave forms and interpreted together without distinction. During application of irregular excitation to the axial piston, the volume decrease of the unsaturated specimens was monitored by the change in water level in the narrow mouth at the top of the inner cell. After the irregularly excited undrained dynamic tests were over, the drainage line was opened to let the excess pore water, if any, drain out of the specimens and additional volume decrease was measured. The volume decrease of unsaturated samples in this phase was less than 10% of the volume reduction monitored during the irregular loading, as shown in Fig. 18c. By adding the volume decreases in these two phases of measurements, the residual volumetric strain, ε_{vr}, was determined.

The dynamic loading tests were performed by changing the amplitude of the irregular excitation in each test, which is expressed by the peak value of axial load.

Thus, the relative magnitude of irregular excitation will be expressed in terms of the maximum stress ratio, $\sigma_{d\,max}/(2\sigma'_o) = \tau_{max}/\sigma'_o$, which is defined as the peak shear stress divided by the initial ambient confining pressure, where $\sigma_{d\,max} = 2\tau_{max}$. During the application of the irregular load, it was possible to measure the maximum shear strain, γ_{max}, which is defined as $\gamma_{max} = (\varepsilon_a - \varepsilon_r)_{max}$, where ε_a and ε_r are axial and radial strain, respectively.

The outcome of the tests is demonstrated in Fig. 19 in a summary from in terms of plots of three variables, that is the dynamic stress ratio, τ_{max}/σ'_o, the maximum shear strain γ_{max} and the residual volumetric strain, ε_{vr}. It is to be noticed here that, as all the specimens had been anisotropically consolidated to the condition of $K_o = 0.7$,

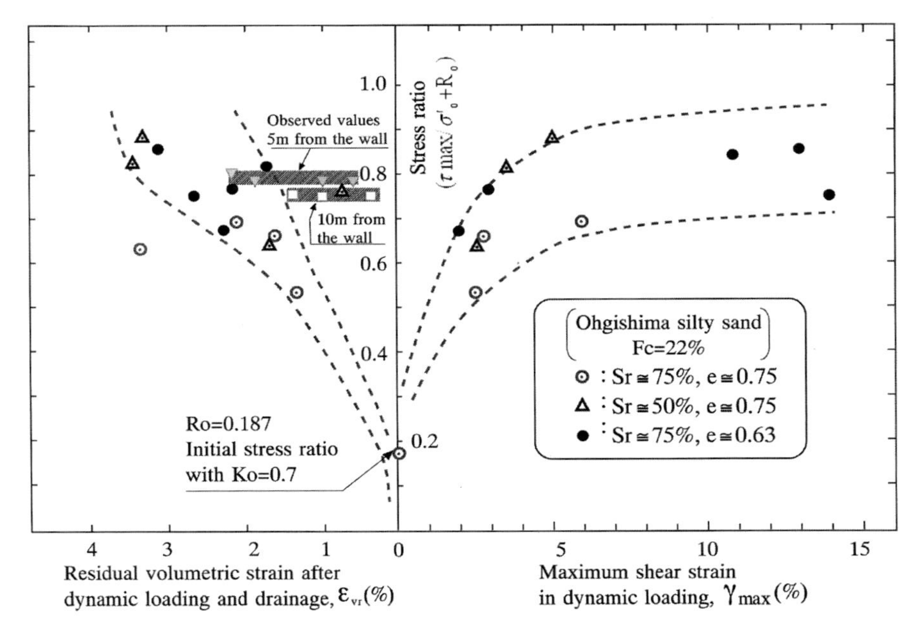

Fig. 19 Cyclic stress versus residual volumetric and maximum shear strains

the samples were subjected statically to an initial shear stress ratio which is equal to $R_o = q/(2p) = 3(1 - K_o)/(1 + 2K_o) = 0.187$, as accordingly indicated in Fig. 19. Under such test conditions, the ordinate in Fig. 19 is a plot of the dynamic stress ratio plus the initial stress ratio, Ro. In this test scheme, no test has been done with $K_o = 1.0$. The reason why tests were not performed with $K_o = 1.0$ was that the tests were intended mainly to study the development of residual shear strains simulating soil conditions under sloping grounds. Even under such a test condition, the results of the tests with $K_o = 0.7$ may be utilized here to a reasonable level of accuracy to draw conclusions necessary for interpretation of the settlements on the level ground such as those observed at the nuclear power station site during the 2007 earthquake. From the results of the tests as summarized in Fig. 19, the following conclusions may be drawn.

(1) In the left-hand side plot of Fig. 19, it can be seen that the residual volumetric strain in the partly saturated silty sand caused by seismic loading tends to increase with increasing dynamic or cyclic stress ratio. However, there seems to be an upper limit of about $\varepsilon_{vr} = 3.5\%$ beyond which the volume decrease can not be produced. Thus, even if the cyclic stress ratio is increased further up corresponding to an increase in the greater intensity of seismic shaking, there would be no further change in the volume of soils in question.

(2) Plotted on the right-hand side of Fig. 19 is the cyclic stress ratio versus the maximum shear strain which the sample underwent during the application of irregular cyclic loading. It can be seen that the value of γ_{max} has a tendency

to increase with increasing cyclic stress ratio, but there is an upper limit of about 0.9 in the cyclic stress ratio beyond which the induced maximum shear strain becomes infinitely large. It is of interest to note that the cyclic stress ratio at which the maximum shear strain γ_{max} becomes infinitely large is nearly coincident with the cyclic stress ratio at which the residual volumetric strain, ε_{vr}, tends to converge to the limiting value of 3.5%.

2.4 Laboratory Tests on Earth Pressure in Repetitive Wall Movements

A series of laboratory tests has been underway at the Tokyo University of Science regarding the gradual increase of earth pressure under repetitive movements of the wall using a sand-filled box with a pin-supported rigid wall H = 50.5 cm in height. (Tatsuoka et al., 2008) The layout of the model test is shown in Fig. 20. The Toyoura sand, the standard sand in Japan, was rained in the box and a deposit of dry sand was formed with a relative density of $D_r = 90\%$. In the tests the top of the pin-supported rigid wall was moved forward by a small amount, and then brought back to the original position, that is, to the state of earth pressure at rest with zero wall movement. This type of forward and backward rotation of the wall around the bottom-hinge was repeated many times and the total lateral force was measured. The pattern of the wall displacement is illustrated in Fig. 20 where the tilt defined as D/H is plotted versus time.

One of the results of the tests in which the wall displacement at the top was 0.3 cm is shown in Fig. 21 where the earth pressure coefficient, Kc, defined as the ratio between the measured total lateral force and the total vertical force $1/2 \cdot \gamma \cdot H^2$

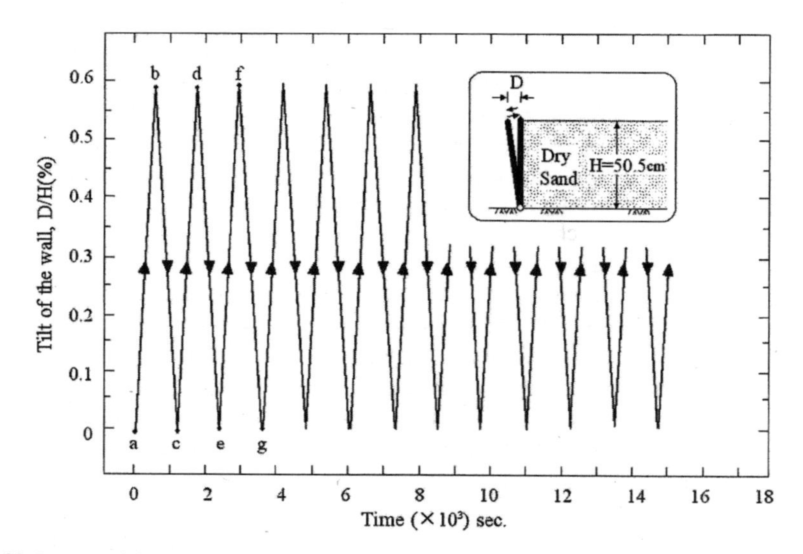

Fig. 20 Layout of the model test and pattern of tilting of the wall

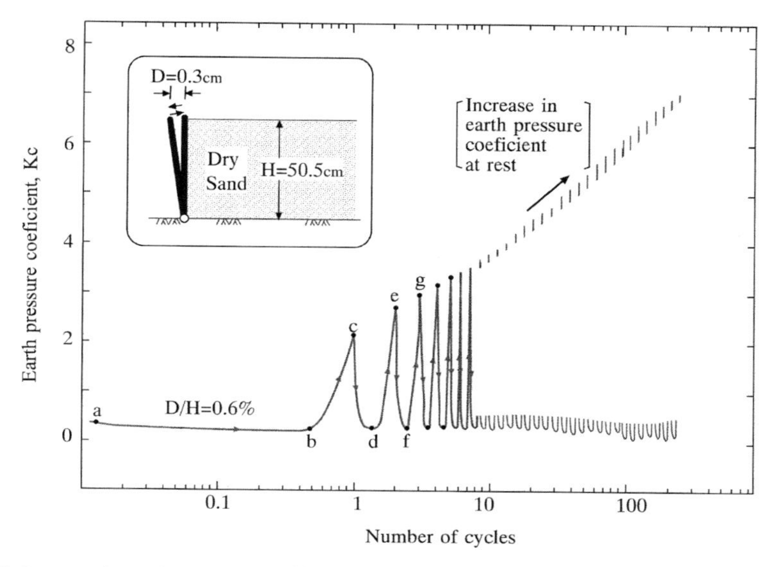

Fig. 21 Increase in earth pressure coefficient with number of cycles (Tatsuoka et al., 2008)

is plotted versus the number of cycles. It is noted that the initial earth pressure coefficient took a value of $K_o = 0.4$ (point a), but it dropped to a value of 0.3 when the wall was tilted forward to produce a state of active earth pressure (point b). When the wall was moved back to the position of zero displacement, the earth pressure is shown to have increased to take a value of $K_c = 2.2$ (point c in Fig. 21). When the wall was tilted forward again, the value of K_c became approximately equal to $K_c = 0.4$ (point d). It is known in Fig. 21 that the repetition of the wall produced gradually increasing earth pressure coefficient at rest as indicated by points c, e and g, while the active earth pressure coefficient practically remained unchanged as shown by points, b, d and f.

If the wall is subjected to the back and forth movement 2–3 times, the result of the test in Fig. 22 indicates that the earth pressure coefficient at rest could take a value of $K_c = 3.0$ which is as much as the coefficient in passive state of earth pressure. The value of increased earth pressure coefficient at rest with increasing number of cycles is plotted in Fig. 22 in a summary form for various tests, each employing different D/H ratios. It is known that, in addition to the increase in the number of cycles, the K_c-value at rest increases also with increasing amplitude of wall displacement. It was commonly observed in the model tests that when the wall is tilted forward, the zone of active earth pressure is generated, but the sand in close proximity to the wall tends to drop into the aperture just behind the wall thereby creating a depression.

When the wall is pushed back to the initial position, the wedge of the aperture-clogging sand is compressed greatly creating a wider deformed zone in the sand fill behind far back to the zone corresponding to the state of passive earth pressure.

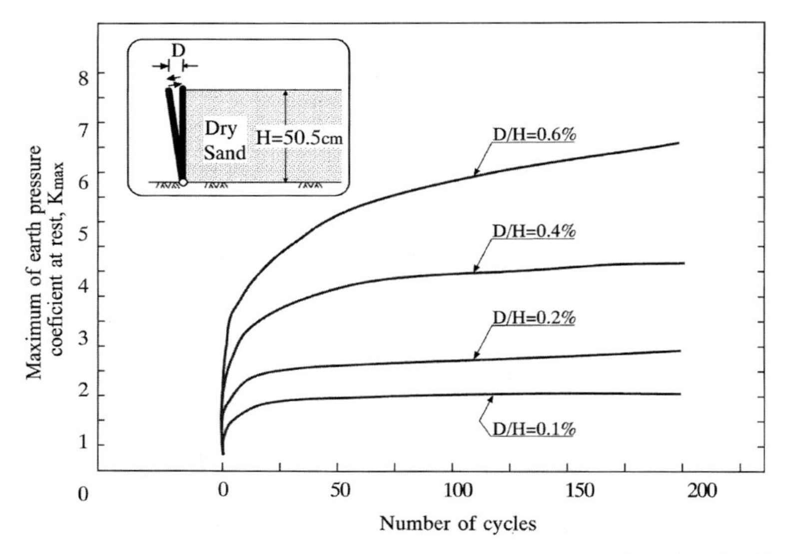

Fig. 22 Increase in earth pressure coefficient at rest with increasing number of cycles (Tatsuoka et al., 2008)

In other words, even if the rigid wall stays at zero deformation, the earth pressure could attain a value as great as, or even greater than that at its passive state. This is the mechanism of ratcheting phenomenon in soil-structure interaction typically observed in the model tests.

2.5 Considerations of the Settlements and Ratcheting Interaction

As mentioned above, there were many places in the premise of the nuclear power stations where settlements occurred in the level ground of unsaturated compacted fills and ratcheting-induced depression in close proximity to the reactor and turbine building. The features of new problems as addressed above will be discussed in the light of the laboratory test data now available as introduced above.

2.5.1 Ground Settlements

Shown in Fig. 23 is a typical cross section in the vicinity of the reactor building No. 1 on the landside where the ground settlements were observed. Details of the location are displayed in Fig. 5. The thickness of the back-fills was 25 m at this place. The silty sand with fines content of about 20 $\sim$ 30% had been compacted very dense. The volumetric strains roughly estimated by dividing the surface settlement by the thickness of the fills were 1 $\sim$ 2.5% for the flat part of the ground, as accordingly indicated in Fig. 23.

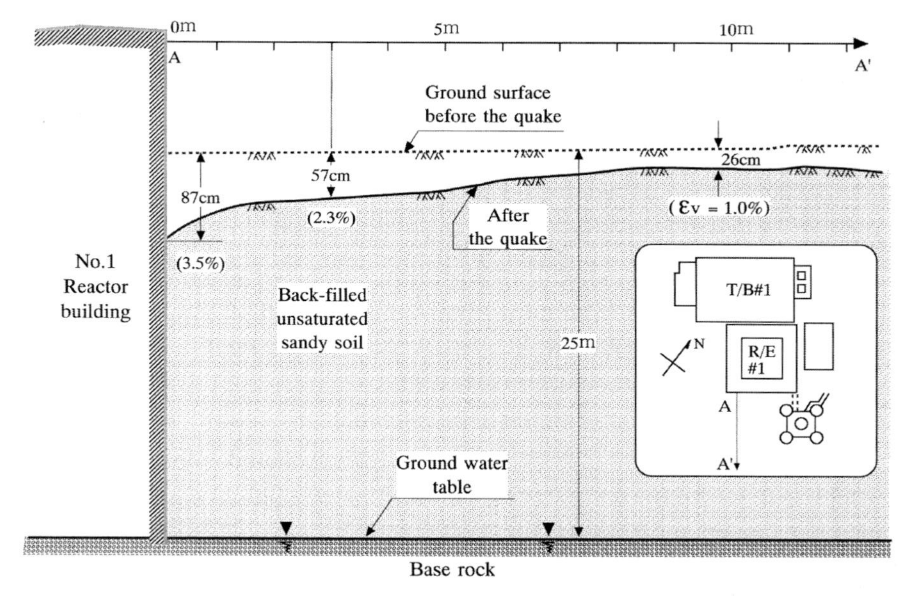

Fig. 23 Settlements and deformations of soil deposits along the cross section A–A′ in the vicinity of No.1 reactor building at the nuclear power station (Tokyo Electric Power Co.)

As introduced above, the peak horizontal accretion recorded at the base of the reactor building No. 1 was on the order of 680 gal and the peak acceleration on the surface recorded at the vertical array site at the Service Hall about 500 m apart was 800 gal. If an absolute value between EW-and NS-components is taken, the peak acceleration was in excess of 1000 gals. With these values it may be possible to estimate the maximum stress ratio by the formulae as follows.

$$\frac{\tau_{max}}{\sigma_v'} = \frac{a_{max}}{g} \cdot r_d \frac{\sigma_v}{\sigma_v'} \tag{1}$$

$$r_d = 1 - 0.015z(m) \tag{2}$$

where the effective vertical stress σ_v' can be taken as equal to the total vertical stress σ_v because of the ground water table maintained at the basement level of the fills. The coefficient r_d is a reduction factor used in the liquefaction analysis. Taking an average depth as z = 12.5 m, Eq. (1) gives the peak stress ratio as follow.

$$\frac{\tau_{max}}{\sigma_v'} = (0.7 - 1.0) \times 0.8125 = 0.57 \sim 0.81 \tag{3}$$

If the silty sand in the nuclear site is assumed to have the grain size characteristics similar to that used in the laboratory triaxial test as described above, the test results demonstrated in Fig. 19, would be utilized to estimate values of volumetric strains. Entering into the diagram in Fig. 19 with the estimated peak stress ratio of 0.57 ∼ 0.81, one can obtain the volumetric strain of the order of 1.5 ∼ 3.5% as determined from the laboratory tests. One of the data on the settlement of the unsaturated compacted soils is shown in Fig. 23 for the reactor building No. 1. The volume decrease of 2.0 and 1.0% at the distance of 5 and 10 m, respectively, is read off from Fig. 23. These data are plotted in Fig. 19 superimposed with the laboratory-obtained data. Similar in-situ data at other locations are also shown in the plot of Fig. 19. With reference to the actually observed values at the nuclear power station site, it may be mentioned that the range of the volumetric strains in partly saturated soils estimated on the basis of the lab. tests would be approximately coincident with those observed in-situ with a reasonable level of credibility.

2.5.2 Earth Pressure Increase by Ratcheting Soil-Structure Interaction

As indicated by the ground sag in Fig. 23 near the building of the No. 1 nuclear power station, the gradual dropping of soils into the aperture between the walls and adjacent ground is considered to have conduced greatly to the increase in the lateral force acting on the walls. As described above, the ratcheting action as such was manifested evidently in the model tests in which the pin-hinged wall was displaced between active and at-rest modes of displacement. In the 2007 earthquake, the major part of the shaking is envisioned to have consisted of three to five cycles as seen in the time histories of acceleration in Fig. 7. It is not clear how much the separation of the walls and the surrounding ground was exactly at the time of the earthquake. However, the separation is envisaged to have penetrated to a depth of about 3.0 ∼ 5.0 m, judging from the observed surface configuration of the near-wall settlement. Thus, it may be roughly assumed that the angle of the walls relative to the surrounding back-filled soils was of the order of $1/25 \sim 3/25 = 4 \sim 12\%$. By referring to the test data in Fig. 22 for the condition of three to five cycles and this tilt angle, it may be roughly assumed that the earth pressure coefficient K_c must have increased to a value well above 5 ∼ 6. However, the walls of the buildings at the nuclear station these were so robust that there was no distress in the structures during the earthquake.

With respect to the high tide-protecting seawalls at Ansei and Kitazono-cho, the forward tilts of the wall are regarded as having resulted from the ratcheting action during the earthquake between the sea walls and back-filled soils. In fact, there were settlements about 50 cm observed in the apron pavement, accompanied by the opening about 1 m wide just behind the wall tilted. Postulating that the walls 7 m high tilted during the main shaking by about 0.5 m the angle of tilt is estimated to have been of the order of $0.5/7.0 = 7\%$ which is sufficiently large to induce increased earth pressure coefficient well over $K_c = 5 \sim 6$. Thus, it may be mentioned with good reasons that even sturdy sea walls defended by stacks of the wave-braking wall had experienced the tilt and probably some forward movement by the great lateral pressure due to three cycles of strong shaking during the earthquake.

On the bases of the discussions as above, it may be more appropriated to consider the coefficient of passive earthpressure in the design of walls which are likely to be subjected to the ratcheting action during the intense earthquake shaking.

3 Concluding Remarks

At the time of the 2007 Niigata Chuetsu-Oki earthquake, multiple suites of acceleration records as strong as 1000 gal were obtained at the basement level of the reactor and turbine buildings in the premise of the nuclear power stations at Kashiwazaki and Kariwa. The destruction by the intense shaking was incurred mainly to the compacted fills of the ground involving overall settlements, differential offsets and large depressions near the buildings. Out of various types of the ground damage, the settlements of unsaturated compacted fills and the ratcheting-induced depression along the periphery of the buildings were taken up in this paper as being new types of ground damage which have not been addressed properly in the past.

With respect to the settlements, it was possible to arrange actual data in a form in which some correlation may be derived between the acceleration-based cyclic stress ratio and the amount of settlements. On the other hands, there was a set of laboratory test data on the residual volume change and the cyclic stress ratio. By comparing the correlation thus obtained in the laboratory samples with the relation derived from the arrangement of in-situ data, it became possible to offer an interpretation to the occurrence of in-situ settlements, on the basis of the laboratory-obtained test data. Thus, it is advisable to make further efforts in the future to come up with some methodology or procedure for estimating the settlements of partly saturated soils undergoing an intense shaking during earthquakes.

Regarding the ratcheting interaction in the vicinity of the buildings, the amount of opening was roughly estimated together with their depth, permitting the amount of tilt of the wall to be evaluated relative to the surrounding soil deposits. On the other hands, there was a set of data from the laboratory model tests indicating earth pressure augmentation resulting from repetitive forward and backward movements of the wall. The laboratory tests indicated evidently that the ratcheting interaction of wall and soil is the generic cause leading to the increased earth pressure. Thus, it was possible to offer an interpretation for unusual increase in lateral thrust during earthquakes.

Another case was cited in this paper from the damage which occurred to seawalls along the coast of Kashiwazaki city during the 2007 earthquake. The increased earth pressure due to the ratcheting interaction was considered as one of the major factors inducing the tilt of the sturdy seawall structure. From the lessons of two cases as described above, it is advisable to advance researches to clarify the mechanism of an extraordinary increase in earth pressure which may be induced as a result of ratcheting action of the wall-soil system subjected to very high intensity of shaking during earthquake.

Acknowledgments In providing this paper, many of the data on the recorded motions and damage features in the nuclear power stations were offered by the Tokyo Electric Power Co. The actual data on the design and damage of the seawalls in Ansei and Kitazono districts were provided by the Nagaoka District Development Bureau through the courtesy of Professor A. Onoue of the Nagaoka Technical College. Arrangements of the recorded accelerations were made by Professors T. Kokusho and T. Ishii of Chuo University in Tokyo. Professor F. Tatsuoka suggested the ratcheting mechanism to be a main factor for the increase earth pressure. The author wishes to express deep thanks to the professionals and organizations as cited above.

References

Nagaoka District Development Bureau, *Report of High Protection walls in Ansei-cho*, Vol.1 (2008).

T. Sakai, Status and Mechanism of Ground Deformation of Kashiwazaki-Kariwa NPS by Niigataken Chuetsu-oki Earthquake, The International Symposium on Seismic Safety of Nuclear Power Plants and Lessons Learned from the Niigataken Chuestu-oki Earthquake (2008).

S. Sawada, Y. Tsukamoto and K. Ishihara, *Residual Deformation Characteristics of Partially Saturated Sandy Soils injected to Seismic Excitation*, Soil Dynamics and Earthquake Engineering, Vol. 26, pp. 175–182. (2006).

F. Tatsuoka, D. Hirawaka, H. Aizawa, H. Nishikiori, R. Soma and Y. Sonoda, *Importance of Strong Connection between Geosynthetic Reinforcement and Facing for GRS Integral Bridge*, Proceeding of the 4th Asian Regional Conference on Geosynthetics, Shanghai, China (2008).

Roles of Civil Engineers for Disaster Mitigation Under Changes of Natural and Social environments and Policies for the Creation of a Safe and Secure Society

Masanori Hamada

Abstract The earth has undergone the drastic environmental changes of global warming, heat island phenomenon in highly urbanized areas, deforestation and reduction in farmland, progressing desertification, and coastal and river erosion over the 20th century. These environmental changes are considered to be the major causes of large-scale typhoons, and hurricanes, drought and abnormally high temperature. Furthermore, our social environment is also changing and it is becoming fragile against natural disasters. This paper gives an overview of the recent natural disasters in the world and associated issues. The author descripes the basic concept of the policies for the natural disasters mitigation and the roles of civil engineers in changing natural environment.

Keywords Natural diasters · Mitigation · Role of civil engineers · Basic policies · Environment

1 Introduction

The natural environmental changes such as global warming, heat island phenomena in mega cities, the decrease of the forest, desertification and erosion of rivers, are resulting in extremely heavy rains and snows, huge typhoons and hurricanes, abnormally high temperature, and high tidal waves. In addition to the change of natural environment, our social environment is also changing and it is becoming fragile against natural disasters. Those are highly congested urban areas, depopulation of rural areas, human habitation on disaster-prone lands, lack of cooperation and communication among recent urban societies, and insufficient infrastructures for disaster mitigation. The characteristics of the natural disasters are changing due to the changes of the natural and social environment.

M. Hamada (✉)
Department of Civil and Environment Engineering, Waseda University, Tokyo, Japan
e-mail: hamada@waseda.jp

A.T. Tankut (ed.), *Earthquakes and Tsunamis,* Geotechnical, Geological, and Earthquake Engineering 11, DOI 10.1007/978-90-481-2399-5_8,

This paper briefly reviews the recent natural disasters in the world and associated subjects. Furthermore, the author discusses the basic concept of the policies for the natural disaster mitigation and the roles of civil engineers.

2 Natural Disasters and Climate Change in the World

2.1 Recent Earthquake and Tsunami Disasters in the World

During recent few years, the disastrous earthquakes and tsunamis have attacked the Asian countries. The 2004 Sumatra earthquake and consequent tsunami killed more than 200,000 people in the areas around the Indian Ocean (Japan Society of Civil Engineers, 2005). In the same year, a devastating earthquake caused serious damage to Niigata Prefecture, Japan (Japan Society of Civil Engineers 2004) due to extensive slope failures in mountainous areas. In 2005, about 70,000 people were killed in Pakistan (Japan Society of Civil Engineers and Architectural Institute of Japan, 2005), and last year, a disastrous earthquake attacked the Java Island, Indonesia.

Figure 1 shows the number of earthquakes and tsunamis with more than 1,000 casualties in each five years period during the last 60 years in the world. The number of events has drastically increased in the last two decades.

The number of earthquakes with magnitudes more than 7.0 and more than 6.0 in the world during the last 60 years is shown in Fig. 2. On the contrary of the increase of the earthquake and tsunami disasters, the number of occurrences of earthquakes with magnitudes more than 7.0 has been decreasing during the last 60 years. The number of the earthquakes with more than 6.0 slightly increased during the last decade, but was not consistent with the rapid increase of the number of the earthquake and tsunami disasters. This suggests that the reason of the increase of the

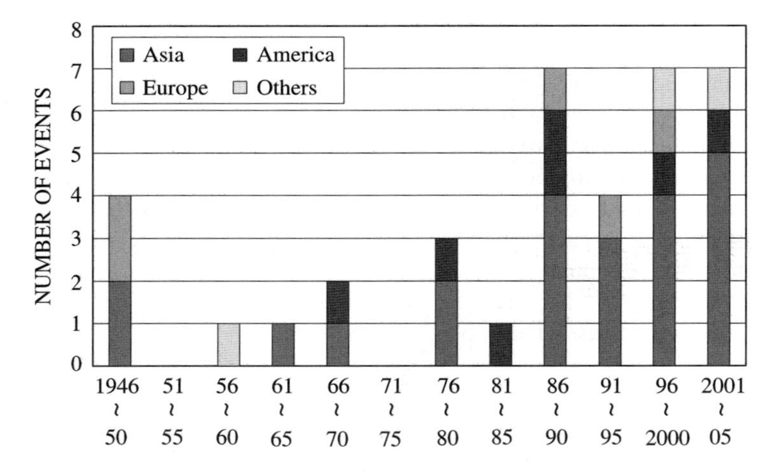

Fig. 1 Damaging earthquakes and tsunamis in the world in the last 60 years (Events with more than 1000 casualties)

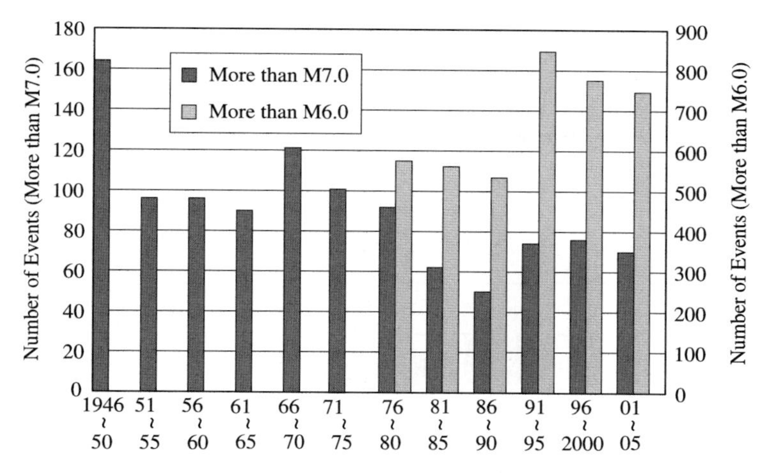

Fig. 2 Number of earthquakes with magnitudes more than 7 and more than 6

earthquake and tsunami disasters is the increase of the vulnerability of our human societies. The regional ratio of the number of the earthquake and tsunami disasters and of the number of causalities during the last two decades is shown in Fig. 3. 25 earthquake and tsunami disasters occurred in the world. Among them, 16 events were in Asian region. About 500,000 people, which are almost 90% of the total number of casualties in the world were killed in Asian region.

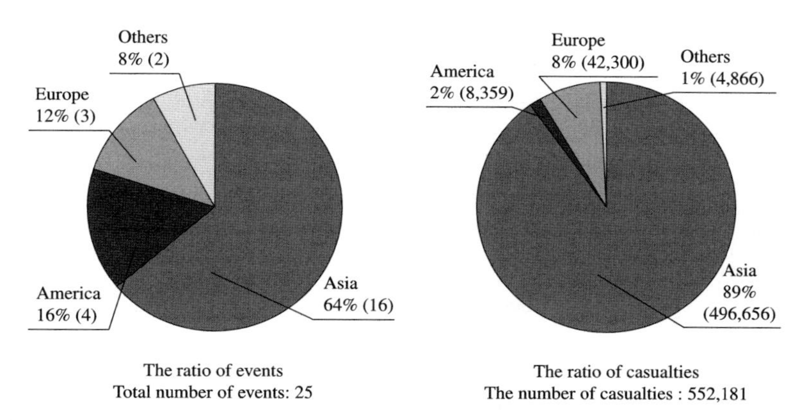

Fig. 3 Regional ratio of earthquake tsunami events and casualties (1956–2005 events with more than 1000 casualties)

2.2 Recent Storm and Flood Disasters in the World

Storm and flood disasters also have resulted in the suffering of the people in the world in recent years. During the last decade, 21 disasters killed 100,000 people. Most of the storm and flood disasters were concentrated in the Asian region and

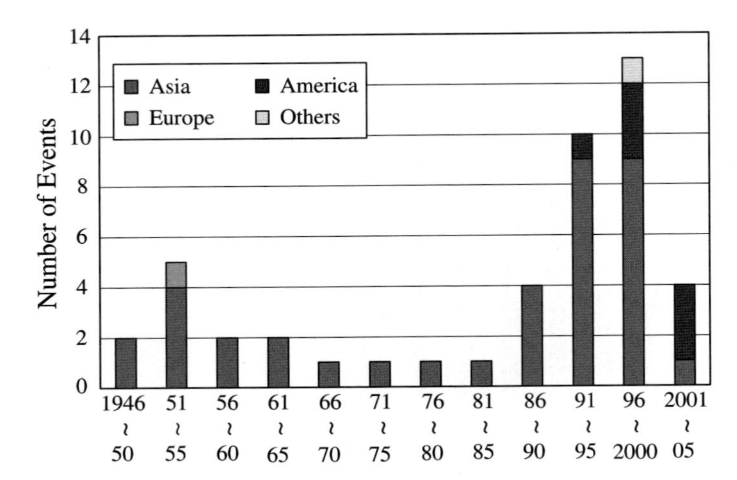

Fig. 4 Damaging storms and floods in the world during the last half century (Events with more than 1000 casualties)

Central America. In 2005, Hurricane Katrina attacked Louisiana in United States, and killed more than 5,000 people. In 2004, the Asian countries such as Japan, Philippines, India and Bangladesh suffered from many typhoons and heave rains, and about 4,000 lives were lost.

Figure 4 shows the number of storm and flood disasters with more than 1,000 casualties in each five years period during the last 60 years. The storm and flood disasters in the world have also increased during the last two decades. And, the disasters in Asia are dominant. The regional ratio of the number of the storm and flood disasters and the number of causalities is shown in Fig. 5. 31 events occurred during the last two decades in the world, and 23 events, which was about three

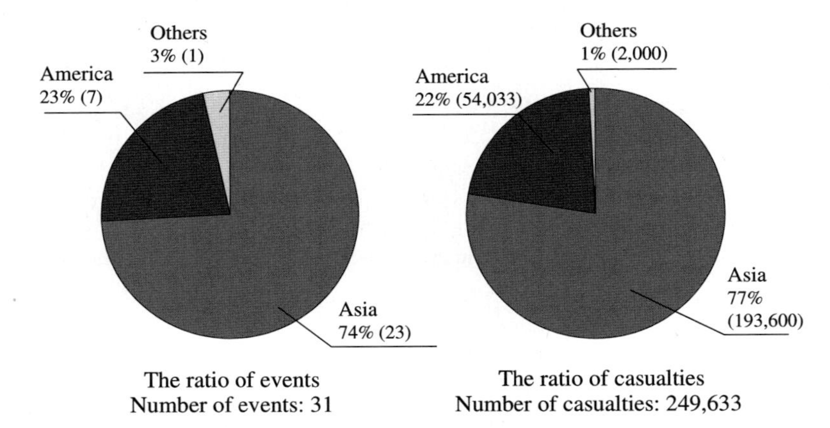

Fig. 5 Regional ratio of storm and flood and associated casualties (1986–2005, events with more than 1000 casualties)

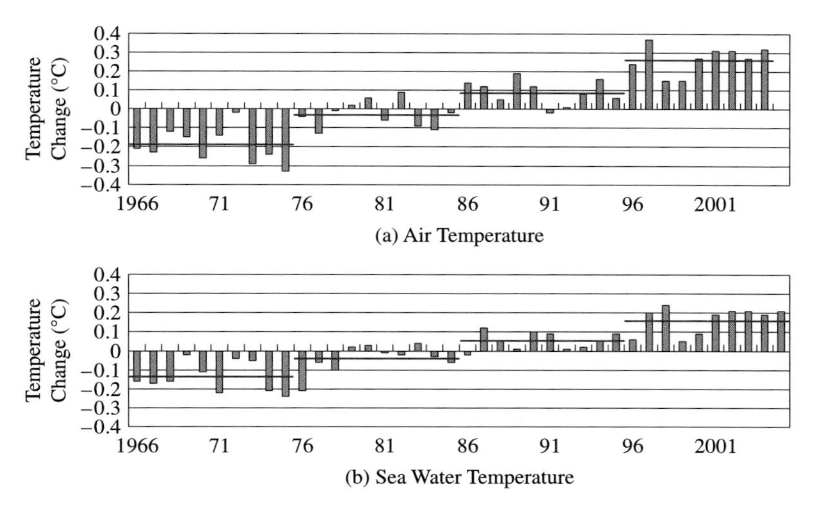

Fig. 6 Temperature change of air and sea water in the world (Japan Meteorological Agency, 2005)

quarters of the total number, attacked the Asian region. About 200,000 lives, which is almost 80% of the total causalities were lost by storm and flood disasters in the Asian region.

There may be two main reasons for the increase of the flood and storm disasters. One is the increase of the vulnerability of the human societies against disasters, which is particularly remarkable in the Asian countries. Another reason may be the global climate change. Figure 6 illustrates the change of air temperature and sea-water temperature in the world during the last 40 years. The air temperature raised about 0.5 degree in centigrade and the sea water temperature raised 0.3 degree. The rise of seawater temperature can be considered to be one of causes of the occurrence of huge typhoons and hurricanes, and high tidal waves.

2.3 Recent Natural Disasters and Climate Change in Japan

The locations of earthquake disasters after the 1995 Kobe earthquake are shown in Fig. 7. The total 6,500 lives were lost, including the Kobe earthquake. We have learned new lessons from each past earthquake. From the Kobe earthquake, we learned lack of earthquake resistance of concrete bridges, subway structures and buildings against the strong earthquake ground motions in the near field of the earthquake fault. A large man-made island reclaimed from the sea was extensively liquefied resulting in severe damage to structures such as storage tanks, and quaywalls.

The Niigata earthquake in 2004 taught us new lessons, one of which was an overlap of different natural disasters. Three days before the earthquake a huge typhoon attacked the same area with heavy rains. The slopes in the mountainous area and

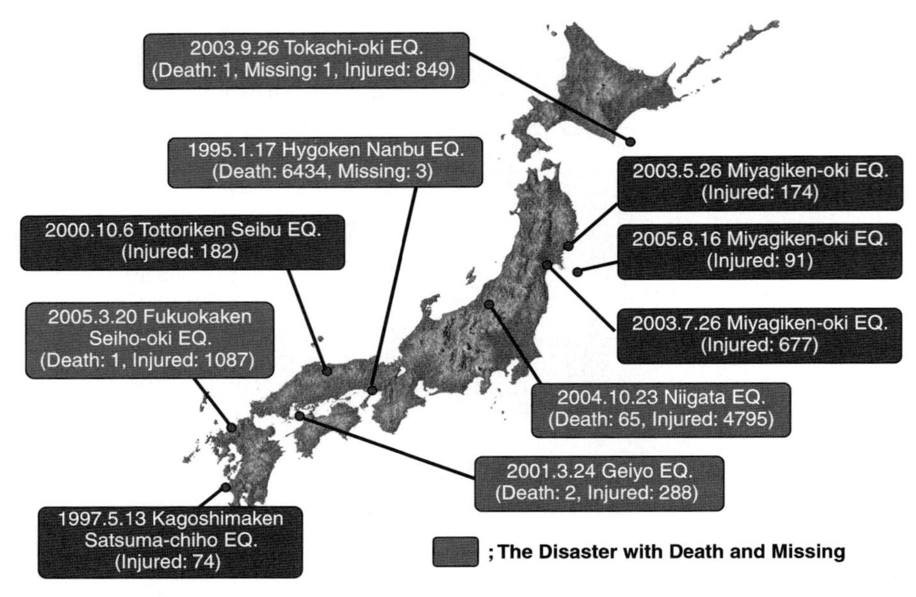

Fig. 7 Recent earthquake disasters in Japan (1995) (The number of death and missing: 6507)

embankments were saturated. The earthquake ground motion caused huge landslides and embankment failures. Liquefaction also caused severe damage to lifeline systems, and strong ground motion in the vicinity of the fault derailed the bullet train, Shinkansen. This derailment of Shinkansen exposed a serious social subject,

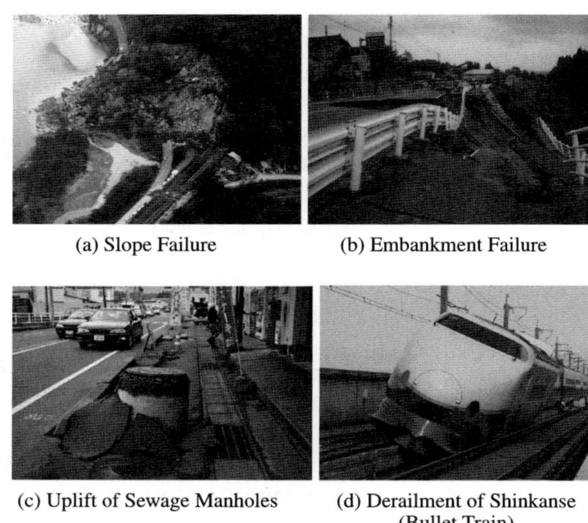

Fig. 8 Disaster by the 2004 Niigata earthquake

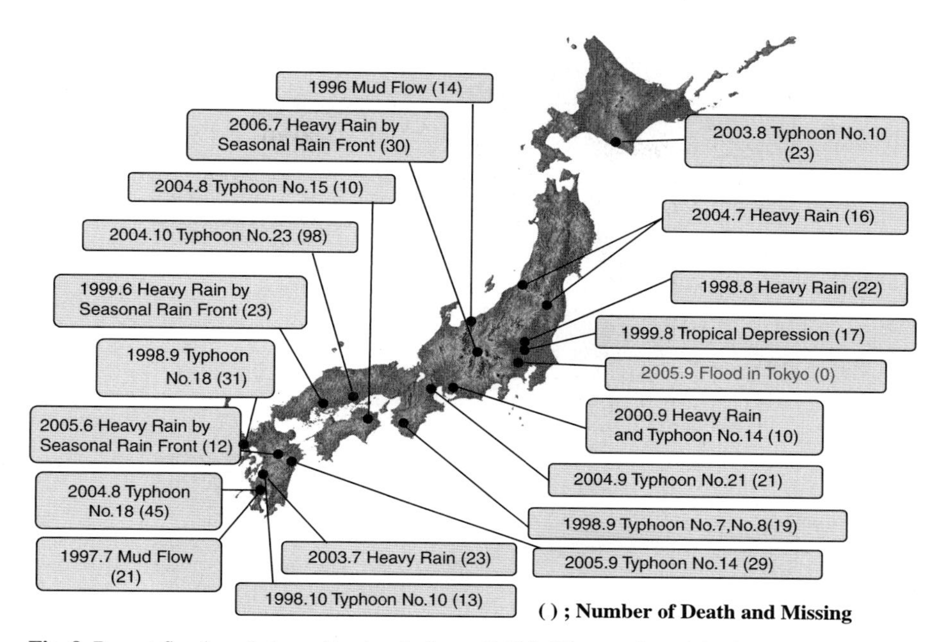

Fig. 9 Recent floods and storm disasters in Japan (1996) (The number of death and missing: 477)

which is how to balance the risk with a large benefit of the high-speed train for the society and the people (Fig. 8).

The recent flood and storm disasters in Japan are shown in Figs. 9 and 10. Flood and storm disasters with casualties occurred twice a year, and about 500 lives were lost during the last decade. Last year, a huge mudflow, which was caused by a heavy rain, attacked residential area in Nagano Prefecture, and killed 12 people. In 2004, Itsukushima shrine, one of national treasures was severely damaged by high tidal wave and strong wind by a large-scale typhoon. In 2005, a downpour with 100 mm rain per hour flooded a wide area of the downtown of Tokyo. This is considered to be caused by the heat island phenomena in highly urbanized mega cities.

Figure 11 shows the change of the temperature and the rain falls in Japan. The mean temperature in whole area of Japan raised about 0.7 °C during the last 30 years, and 1.0 °C in Tokyo. The number of the rainfalls with more than 100 mm per hour in Japan has been increasing during the last decade.

3 Basic Concepts of the Policy Against Future Disasters Under the Change of Natural and Social Environments

The natural environment is changing (Fig.12). Those changes are global warning, heat Island phenomena in urbanized area, deforestation, desertification and erosion of river and seashore. The change of natural environment is increasing natural

(a) 2006 Mud Flow by Heavy Rain (Nagano)

(b) 2004 Itsukushima-shrine which was Destroyed by Typhoon No.18 (Hiroshima)

(c) 1999 Tidal Wave by Typhoon No.18 (Kumamoto)

(d) 2005 Flood by Heavy Rain (Tokyo)

Fig. 10 Recent Flood and storm disaster in Japan

disasters. Those are extremely heavy rains and snows, huge typhoons, hurricanes and cyclones, drought, abnormally high temperature and high tidal waves due to the rising of sea water level. In addition to the change of natural environment, our social environment is also changing, becoming fragile against natural disasters. Those are too congested urban areas, depopulation of rural area, human habitations on fragile ground, lack of cooperation and communication among the recent urban societies, budget deficit of central and rural governments, and finally poverty. The poverty is the most important factor for the increase of the natural disasters in the Asian countries. The poverty is expanding the disaster, and the disaster is worsening the poverty.

The characteristics of natural disaster are changing due to the change of natural and social environments. What is the basic policy against these kind of natural disasters? The key point for the measures against future disasters is how to prepare unexpected natural phenomena and against external forces largely exceeding the design level. In another word, how to prepare against natural disasters with a huge scale, but low probability of occurrence.

Figure 13 illustrates a basic concept for the measures against huge natural disasters with low probability of occurrence. That is a combination of hardware measures and software measures. Hardware measures mean, for an example, reinforcement

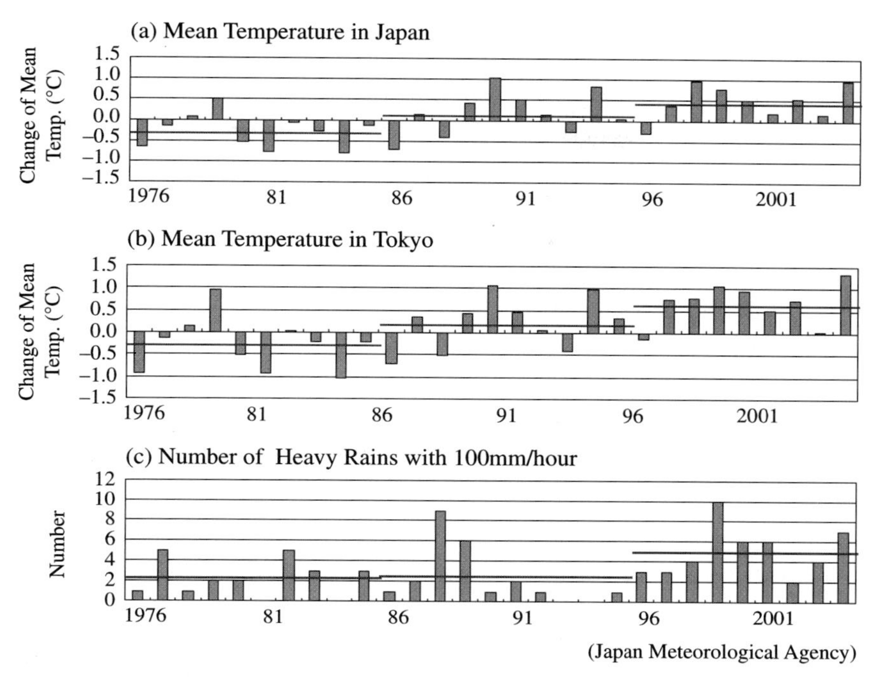

Fig. 11 Change of Mean temperature and rain falls in Japan (Japan Meteorological Agency, 2005) (1976–2004)

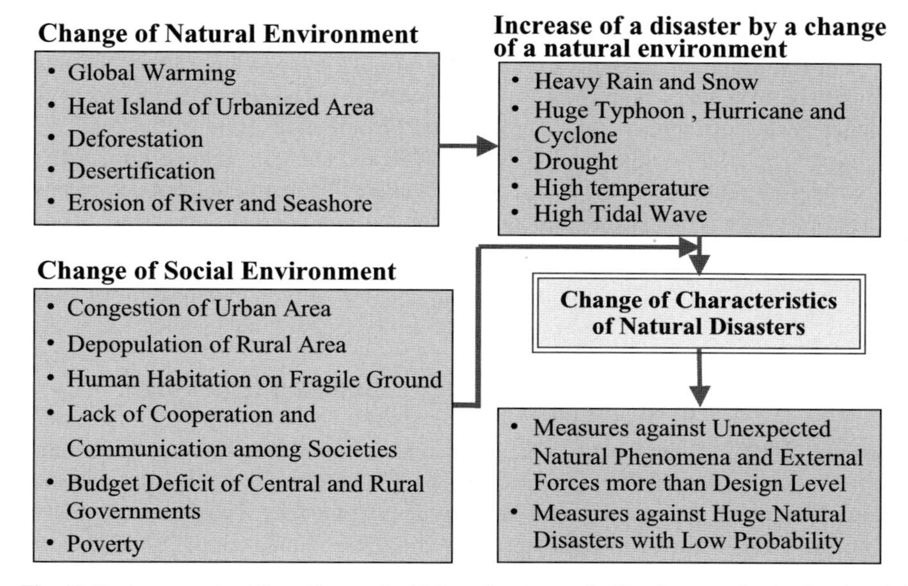

Fig. 12 Basic concepts of the policy against future disasters under the changes of natural and social environment

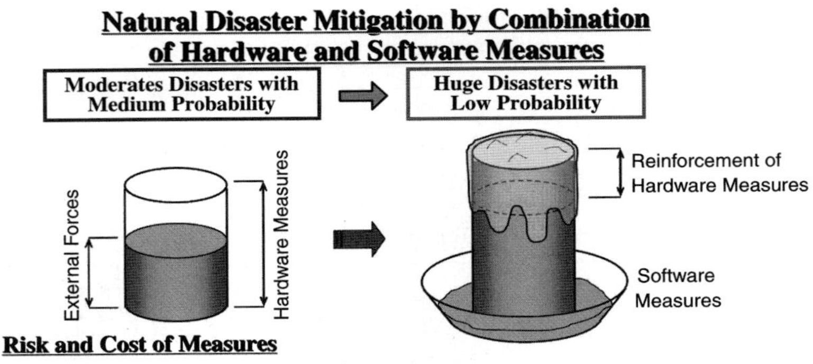

Risk and Cost of Measures

Risk (Total Loss × Probability) ⋚ Cost of Measures

Factors for Estimation of Total Loss

Loss of Human Lives and Properties , Decline of National Power , Ruin of National Land , Psychological Damage to People

Consensus of People on Investment for Measures

Evaluation and Public Release of Probable Risk , Common understandings of Responsibilities for Public Measures , Community Cooperation and Private Efforts

Fig. 13 Measures against huge natural disasters with low probability

of dikes against floods, and the soft ware measures are evacuation systems during flood and the education of people. Moderate disasters with medium probability are prevented mainly by hardware measures. However, against huge disasters with low probability, the disaster is reduced both by hardware and software measures.

The problem is how to determine the rational level of the investment for disaster mitigation. One of methods to judge the rational level of the investment is the comparison of the risk with the cost of the measures. The risk is estimated as the product of the total loss with the probability. For the estimation of the total loss, we have to take into consideration, various factors, not only loss of human lives and properties, but also probable national power decline resulting from the disaster, ruining of national landscape and furthermore, psychological damage to people. And the consensus among the people is essential to determine the rational level of disaster mitigation measures.

4 Future Earthquakes Threatening Japanese Mega Cities

As shown in Fig. 14, large earthquakes, which have high probabilities of occurrence, are threatening the Japanese mega cities such as Tokyo, Nagoya, Osaka and Sendai. Figure 14 shows the source areas of those future earthquakes. The three earthquakes along the Nankai trough in the Pacific Ocean, Tokai, Tonankai and Nankai earthquakes have large magnitudes of 8 or more. The probabilities of the occurrence of these earthquakes are predicted to be very high. In the case of Tokai earthquake, the probability of occurrence in the next thirty years is more than 80%.

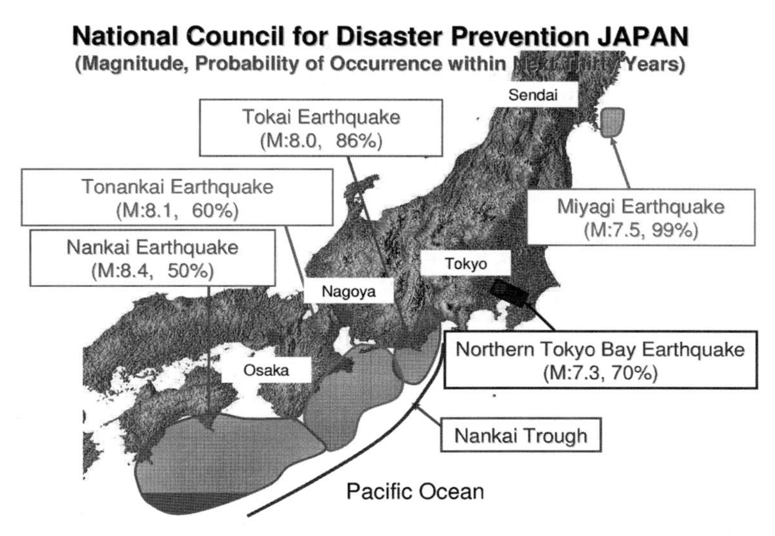

Fig. 14 Future earthquakes with high probabilities to hit the mega cities in Japan (Cabinet office, Government of Japan, 2006)

The northern Tokyo Bay earthquake with a magnitude of 7.3 has also high probability, 70% within next thirty years. In the case of Miyagi earthquake with a magnitude of 7.5, which is threatening the Sendai City, the probability is estimated as 99%. As shown in the figure, many mega cities of Japan are under the threat of being hit by disastrous earthquakes in very near future.

Table 1 shows the result of the damage assessment by the Northern Tokyo Bay earthquake by the National Council of Disaster Prevention of Japan under a condition, that is, the wind speed is 15m/s. If the earthquake did occur at 6 PM, the

Table 1 Assessment of damage by the Northern Tokyo bay earthquake (Cabinet office, Government of Japan, 2006)

- Tokyo and The Neighboring Three Prefectures- (Wind Speed 15 m/s)				
Items		5:00 AM	6:00 PM	Kobe Earthquake
Number of Collapsed or Burnt House and Buildings	Ground motion	150,000	150,000	110,000
	Fire	160,000	650,000	7,000
	Liquefaction, Slope sliding, etc	35,000	50,000	46
	Total	360,000	850,000	117,000
Death	Collapse of houses	4,200	3,100	4,915
	Fire	400	6,200	550
	Slope sliding	1,000	900	37
	Total	5,600	11,000	5,520
Number of People unable to go home (12:00AM)			6,500,000	
Refugees			4,600,00	237,000
Economic Loss			$ 1.1 Trillion	$ 110 Billion

Council estimated the number of collapsed or burnt houses and building as about 850,000 in Tokyo area, including the neighboring three prefectures, that would be about seven times of the damage caused by the Kobe earthquake. The resulting fires would burn more than 650,000 houses and buildings. The total number of the death is estimated as 11,000, and among them about 6,000 people would be killed by the fire after the earthquake. Furthermore, if the earthquake occurred at noon, 6.5 million people could not go home on the day of occurrence of the earthquake due to the destructive damage to the transportation systems. And the number of refugees is estimated to be 4.6 million, which is about 20 times of the number in the case of the Kobe earthquake. The total economic loss reaches over 1.1 trillion US dollars, which is almost 1.5 times of the total amount of national budget of Japan.

5 Roles of Civil Engineers for Natural Disaster Mitigation

The first role of civil engineers for the natural disaster mitigation is the development of technologies for enhancement of infrastructures, such as technologies for the improvement of soft soil, high-performance structures, and warning and rescue systems. The second role is actual construction of infrastructures with high natural disaster resistance. The third role of civil engineers is the involvement in the rescue operation, and restoration and reconstruction works after natural disasters (Fig. 15).

A large number of infrastructures such as highways, railways, port-harbor facilities have been constructed on improved soft alluvial and man-made ground along the seaside of the Tokyo Bay. Various kinds of methods for improvement of soft alluvial and artificial ground have been developed. Technologies such as passive and active control systems have been developed to reduce the dynamic effects on

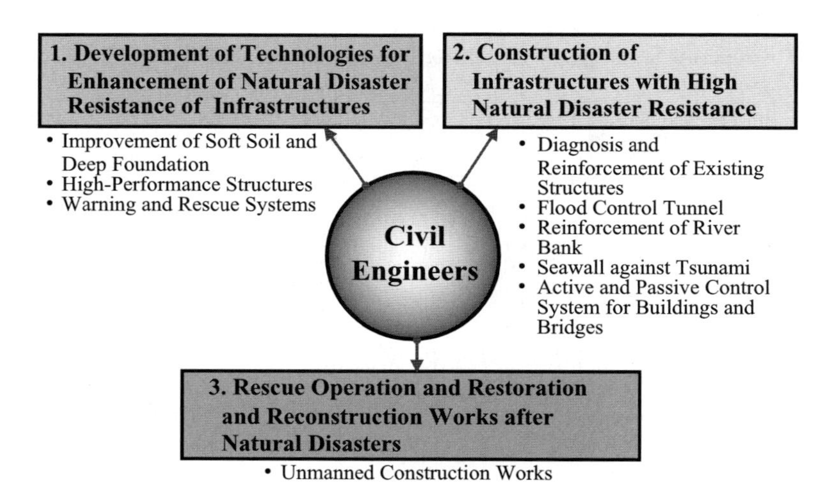

Fig. 15 Roles of civil engineers for natural disaster mitigation

buildings and bridges during earthquakes, and have been applied to a large number of structures.

The 1995 Kobe earthquake destroyed a huge amount of infrastructures. Therefore, Reinforcement of existing infrastructures against future earthquakes has been one of most important roles of civil engineers after the Kobe earthquake. A numerous number of infrastructures such as railway and highway bridges, and subway stations have been reinforced.

The real time earthquake warning system has been developed and applied to some practical uses as shown in Fig. 16. In this system, the magnitude and the epicenter of earthquakes will be judged by the ground motion records in the vicinity of earthquake faults, and if the earthquake has such a large magnitude as to cause serious damage, the warning will be sent to various organizations before the arrival of the main ground motion of the earthquake. Based on the warning, the high-speed trains will be stopped, and the road and air traffic will be carefully controlled. The operation of the various kinds of plants will be shutdown, and the water gates will be closed against the tsunami. In the case of Tokai earthquake, the time allowance before arrival of the main shocks is estimated as about 50 s in Tokyo area.

In low land areas in Tokyo, which has lower elevation than river water level, so-called "super levee" has been constructed against future earthquakes and for development of the areas with high natural disaster resistance, as shown in Fig. 17. New banking behind the original riverbank elevated a wide area and the area was redeveloped to enhance the natural disaster resistance, by creating open spaces for the rescue operation and by constructing high earthquake resistant buildings.

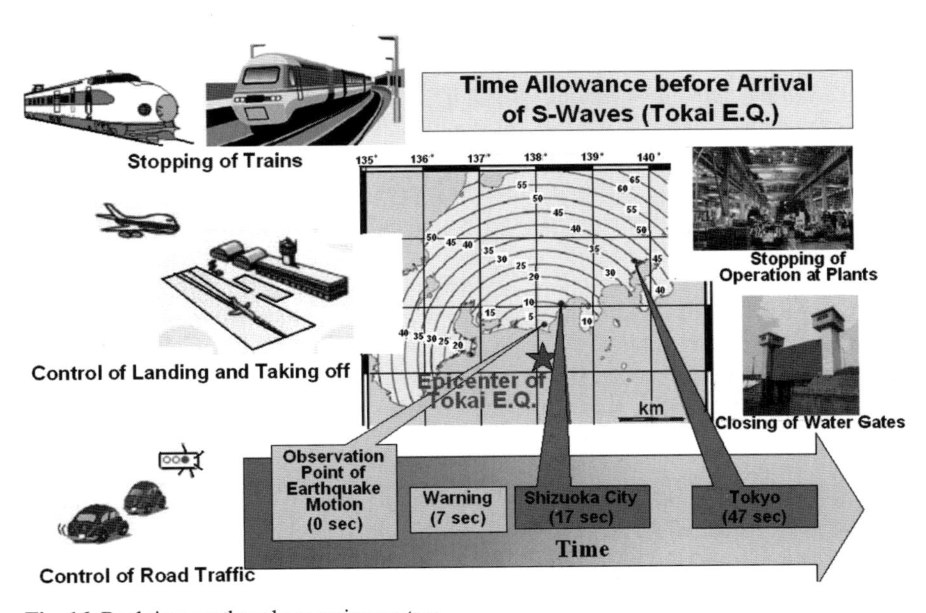

Fig. 16 Real time earthquake warning system

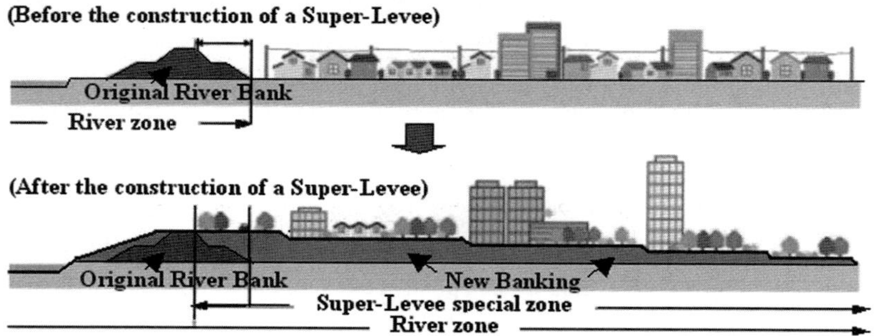

Fig. 17 Construction of the Super-Levee in Low Land Area of Tokyo against Floods and Earthquakes

6 Proposals for Policies and Measures

(1) *Paradigm Shift for Creation of a Safe and Secure Society*: Against a background in which the natural environment is changing, and land utilization and social systems are becoming vulnerable, the government should make a paradigm shift in policies for natural disaster mitigation, away from a short-term perspective focusing on economic growth and towards the long-term creation of a safe and secure society.

(2) *Development of Infrastructure*: Tax revenues should be properly allocated for infrastructure development focusing on natural disaster mitigation. In determining an appropriate level of investments in infrastructure development, the impacts of future natural disasters should be assessed and evaluated by taking into consideration not only the loss of lives and properties but also decline in the nation's power, degradation of landscape and psychological damage to the people.

(3) *Rearrangement of National Land Utilization*: National land utilization should be more balanced for natural disaster mitigation. To achieve that, the following measures should be realized: decentralizing the population and assets from the mega cities, movement of residence of people from disaster-vulnerable zones to resilient areas, establishing backup systems for maintaining the function of the capital, Tokyo, and developing transportation systems for recovery and reconstruction activities.

(4) *Application of "Hardware" and "Software" Countermeasures*: For successful mitigation of large-scale natural disasters, in addition to "hardware" countermeasures like reinforcement of buildings, houses, levees and embankments, "software" countermeasures should be substantially developed such as disaster education, dissemination of disaster-related knowledge and experience, development of comprehensive systems for emergency information, evacuation and rescue, and medical treatment.

(5) *Assessment of Vulnerability and Potential Risks of Depopulated Areas*: Isolated islands, coastal zones and mountainous areas have become vulnerable to natural disasters due to decreasing population and aged nuclear families. The vulnerability of those areas should be assessed for emergency relief and rescue operations.

(6) *Establishment of Integrated Disaster Management Policies and Strategies by Central and Municipal Governments*: Each ministry and agency involved with natural disaster mitigation should establish and implement integrated disaster management policies and strategies in mutual cooperation. In the meantime, municipalities should establish disaster management systems and countermeasures, and then build cooperation and coordination with their neighboring municipalities. For the municipalities' efforts, the central government should provide financial and other support. In the event of a large-scale disaster, which may cause serious impact on wide areas of the country, the central government should take the initiative in operating rescue, restoration and reconstruction activities.

(7) *Development of Public Awareness of Disaster Risks and Preparedness*: Hazard maps should be prepared for public reference. Through this effort, public awareness and concern regarding natural disaster mitigation can be enhanced. The government should also educate the public to acquire proper disaster awareness by showing the nation's vulnerability to natural disasters following changes to social environment such as the aging society and low birthrate, the increase of nuclear families highly computerized daily life and internationalization of society. In collaboration with the public and municipalities, the government should establish a "disaster-aware society" which will be able to respond resiliently to the attacks of natural disasters.

(8) *Education on Natural Disaster Mitigation*: Education should be strongly promoted in teaching geography, geology and other subjects related to natural disasters in schools so that younger generations can acquire fundamental knowledge of natural phenomena and the mechanisms of disaster occurrence, and obtain the proper understanding and judgment to prepare for disasters.

(9) *Promotion of Activities by NPO and NGO*: In nationwide activities, natural disaster mitigation should be promoted in an appropriate combination of public works, community collaboration and individual efforts. Non-Profit Organizations (NPOs) and Non-Governmental Organizations (NGOs) have an important role in strengthening community collaboration through disseminating disaster knowledge and experience, and participating in emergency relief activities. The central government and regional public agencies should promote the development of NPOs and NGOs by financial and logistics support.

(10) *International Cooperation in Disaster Mitigation*: Japan, which has achieved rapid economic growth while dealing with natural disasters, has been requested to share its experience and knowledge of disaster mitigation with the natural disaster-prone countries in the Asian region in particular. Japan should make efforts to respond to the requests of these countries, strengthening its international cooperation in disaster mitigation activities.
The disaster mitigation activities should be operated from a multidisciplinary perspective including sociology, economics, agriculture, environment studies, science, engineering, and education. Accordingly, it is essential to establish close collaboration and coordination among the concerned ministries and agencies, and they should build a strategic framework for international assistance and cooperation beyond borders.

(11) *Sustainable Systems and Frameworks for Natural Disaster Mitigation*: Sustainable systems and frameworks should be built for the realization of appropriate disaster mitigation measures. That can be done by continually assessing the vulnerabilities of land utilization and social systems, and developing infrastructure for disaster mitigation.

(12) *Improvement of Natural Phenomena Observation and Monitoring Systems*: Natural phenomena observation and monitoring systems should be continuously updated so that any sign can be identified and followed to predict large-scale earthquakes, tsunamis and volcanic eruptions. The impact and characteristics of large-scale disasters, which may occur once in several hundreds or even thousands of years should be studied by geological and geophysical surveys.

(13) *Development of Numerical Models for the Prediction of Climate Change*: The causes of global climate change and warming should be clarified in relation to natural environmental variations and human activities. In addition, more accurate predictions should be attained by utilizing the data obtained from satellite monitoring and the results of computer simulations. In the meantime, uncertainties in the predictions should be identified and taken into account in the planning of disaster mitigation measures.

(14) *Promotion of Research and Development on Improvement of National Land Utilization and Social Systems against Natural Disasters*: To overcome the disaster vulnerabilities of national land utilization plan and social systems, public and private research institutes and universities are required to collaborate in conducting comprehensive researches and surveys. The government should provide them with its organizational, administrative and financial support.

(15) *Dissemination of Research and Study Results and Development of Human Resources*: Public and private research institutes and universities are requested to release their research results in forms accessible to the public and disaster-concerned organizations, and to take initiatives to promote human resource development and natural disaster education nationally and internationally.

(16) *Promotion of Multidisciplinary Researches and International Collaborations*: The Science Council of Japan (SCJ) is requested to present proposals for policymaking and promotion of researches on disaster mitigation. Additionally, the SCJ is expected to promote multidisciplinary, inter-organizational researches in the fields of science, engineering, bioscience, social science and humanities, and to transfer technology and knowledge for disaster mitigation across borders through promotion of international joint research programs.

7 Conclusions

Natural disasters such as earthquakes and tsunami, and storm and flood disasters have been increasing during last two decade, particularly in the Asian countries. Furthermore, due to the change of natural and social environments, the scale of the natural disasters is expanding. For the mitigation of the future natural disasters, the roles and the responsibilities of civil engineers are increasing. Furthermore, the cooperation and collaboration of civil engineers in the world are essential to accomplish these errands.

JSCE and the other organizations of civil engineers in the world have been in a good partnership during decades, and have served people for the safe living and welfare. However, the recent natural disasters in the world taught us again that the development and application of technologies and knowledge for natural disaster mitigation are the urgent subject for civil engineers. We have to strengthen and to further advance the cooperation between two societies.

References

Cabinet Office, Government of Japan: White Paper on Disaster Prevention, 2006.

Japan Meteorological Agency: A Report on Abnormal Weather 2005, 2005.10.

Japan Society of Civil Engineers: The Damage Induced by Sumatra Earthquake and Associated Tsunami of December 26, 2004, 2005.8.

Japan Society of Civil Engineers: General Investigation about the Damage of Social Infrastructure Systems due to the 2004 Niigata Earthquake/Result of Investigations and Urgent Proposals, 2004.11.

Japan Society of Civil Engineers and Architectural Institute of Japan: A Quick Report on Kashmir Earthquake, 2005.11.

Microzonation for Urban Planning

Atilla Ansal, Gökçe Tönük, and Aslı Kurtuluş

Abstract Microzonation is identification of areas having different earthquake hazard potentials and will primarily serve for urban planning and land use management. The two principal factors controlling earthquake loss are site response and structural features. The seismic microzonation maps would indicate the distribution of site response with respect to ground shaking intensity, liquefaction and landslide susceptibility; thus providing an input for urban planning and earthquake mitigation priorities at an urban scale. It is also possible to estimate building damage and causalities based on microzonation maps used as an input to earthquake risk scenarios. These estimates may be very approximate based on the accuracy of the input data and methods of analysis. However, they can also be more realistic when more comprehensive data and more sophisticated analysis methods are implemented. The first stage in an earthquake risk scenario is the estimation of the earthquake hazard on the ground surface based on seismic microzonation maps for detailed assessment of earthquake response on the ground surface. The seismic microzonation maps, in other words earthquake hazard maps, are prepared using estimated earthquake characteristics on the ground surface based on local site conditions and input earthquake characteristics. The estimation of damage and causalities for buildings, lifelines and transportation networks is the second stage and is considered as earthquake damage scenarios. In order to assess the effects of earthquakes in urban areas, it would be necessary to compile historical earthquakes, geological, geotechnical and seismological data, to evaluate the probabilistic and deterministic earthquake hazard, to determine the variation of earthquake characteristics with respect to local site conditions, and microzonation and damage maps that need to be drafted utilizing GIS software packages. Within the contents of this chapter, the first stage of earthquake risk scenarios, microzonation – probabilistic earthquake hazard scenario will be presented based on some case studies.

Keywords Microzonation · Earthquake scenarios · Site effects · Site response analysis

---

A. Ansal (✉)
Kandilli Observatory and Earthquake Research Institute, Boğaziçi University, Istanbul, Turkey
e-mail: ansal@boun.edu.tr

A.T. Tankut (ed.), *Earthquakes and Tsunamis,* Geotechnical, Geological, and Earthquake Engineering 11, DOI 10.1007/978-90-481-2399-5_9,

1 Introduction

The earthquake hazard is spatially distributed in relation to earthquake sources (faults) and local geological site conditions. Mapping the variation in earthquake hazard at an urban scale makes it possible to select relatively less affected sites for the allocation of appropriate land use. Urban development patterns can be oriented toward these relatively less affected zones to minimize possible earthquake damages.

Seismic microzonation maps provide a detailed assessment of potential earthquake effects, which can provide guidance in urban planning and development. At the urban level identification of relative hazard variations due to differentiated earthquake characteristics can be used to introduce earthquake effects as a factor in the urban development and land use decisions. Microzonation or more specifically the variation of earthquake characteristics at an urban level is also essential for the structural designer and builder to enable them to anticipate problems related to amplification of ground shaking, liquefaction and landslide susceptibilities. However, site-specific investigations would still be required for design of special buildings and for rehabilitation and retrofit projects.

Microzonation would encompass the variations in earthquake hazard parameters; e.g. surface faulting and tectonic deformation; ground shaking intensity; liquefaction, ground spreading and settlement susceptibility; slope stability problems like landslides or rock falls; and earthquake-related flooding due to tsunamis. In order to assess the above-mentioned effects for a region selected for microzonation, sufficiently detailed seismological, geophysical, geological, and geotechnical investigations are considered necessary.

Site specific free field earthquake characteristics on the ground surface are the essential components for microzonation with respect to ground shaking intensity, liquefaction susceptibility and for the assessment of the seismic vulnerability of the urban environment. The adopted microzonation methodology is based on a grid (cell) system and is composed of three stages: In the first stage, regional seismic hazard analyses need to be conducted to estimate earthquake characteristics on the rock outcrop for each cell. The probabilistic seismic hazard analysis should be at the regional level to determine the earthquake characteristics for 475 years return period corresponding to 10% exceedance in 50 years. In the second stage, the representative site profiles should be modeled based on the available borings and in-situ tests. The third stage involves site response analyses for estimating the earthquake characteristics on the ground surface and the interpretation of the results for microzonation (Ansal et al., 2004b). In addition to the generation of base maps for urban planning, microzonation with respect to spectral accelerations, peak acceleration and peak velocity on the ground surface can be used to assess the vulnerability of the building stock (Ansal et al., 2009, 2007a) and lifeline systems (Ansal et al., 2008).

The spectral accelerations on the ground surface to be used in the vulnerability assessment of the building stock are determined based on elastic acceleration response spectra obtained from site response analyses. The procedure proposed by Ansal et al. (2005a) is used to determine short period ($T = 0.2\,s$) and long period

(T = 1 s) spectral accelerations by fitting the best NEHRP envelop spectra to the average elastic acceleration response spectra obtained for each cell.

The adopted microzonation methodology was developed based on microzonation studies conducted in Turkey during the last decade. The initial applications of adopted microzonation methodology were conducted for Istanbul Bağcilar (Ansal, 2002) and Silivri (Ansal et al., 2004a). An important improvement in the methodology was achieved during the DRM project and related pilot studies for Adapazarı and Gölcük towns after 1999 Kocaeli and Düzce earthquakes (Ansal et al., 2004c, b; Studer and Ansal, 2004).

The developed microzonation methodology was later applied to Zeytinburnu Municipality as a pilot project for Istanbul Earthquake Master Plan (Kılıç et al., 2006; Ansal et al., 2005a) and to six municipalities (Bandırma, Bakırköy, Eskişehir Gemlik, Körfez ve Tekirdağ) in the World Bank project MEER (Ansal et al., 2007b, 2006a). During this period also a Microzonation Code was drafted for Istanbul Metropolitan Municipality (Ansal et al., 2005b). The methodology was farther developed during EU FP6 LessLoss project (Ansal and Tönük, 2007b; Ansal et al., 2007a, Spence, 2007). And most recently a microzonation study was conducted for Bolu (Ansal et al., 2007c).

2 Seismic Hazard Analysis

The first stage in the microzonation studies is the estimation of the regional earthquake hazard (e.g. Erdik et al., 2004). Earthquake characteristics to be used in a microzonation study are evaluated based on regional tectonics and seismic activity. In the case of microzonation for urban planning, it is preferable to adopt probabilistic earthquake hazard assessment, since the purpose is to provide general guidelines for land use and urban planning. By definition, probabilistic approach encompasses all possible and most unfavorable earthquake ground motion characteristics that may be experienced at a site with respect to exceedance probabilities for a given time period and return periods.

The other option for the regional earthquake hazard is to use simulated acceleration time histories generated using deterministic fault rupture models (Ansal et al., 2009). Earthquake hazard results differ drastically depending on the approach whether it is probabilistic or deterministic. The damage level at various locations of the investigated city may differ drastically based on selected fault rupture models in deterministic scenarios due to directivity effects. The probabilistic approach that takes into account all these ambiguities arising from the earthquake rupture models is more suitable for microzonation and urban planning purpose.

In seismic hazard analysis, the variation of spectral accelerations at least for two spectral periods preferably one for short and one for long period ranges should be calculated using appropriate attenuation relationships considering regional source and site properties. Generally, the results of the earthquake hazard analysis corresponding to 475 year return period (10% exceedance probability in 50 years) for the engineering bedrock (Vs $\geq$ 750 m/s) are determined in terms of peak ground and spectral accelerations at T = 0.2 s and T = 1 s for the selection and scaling of input motion time histories for site response analysis.

Independent of the methodology adopted for the earthquake hazard assessment, whether it is probabilistic or deterministic, hazard compatible recorded or simulated acceleration time histories are used to conduct site response analyses for the investigated area.

All available previously recorded acceleration time histories compatible with the earthquake hazard assessment in terms of probable magnitude, distance and fault mechanism could be selected as input outcrop motion. It is preferable to conduct large number of site response analyses using different input acceleration time histories to eliminate the differences that are observed between different sets (Ansal and Tönük, 2007a, b) and also to take into account the variability due to the earthquake source characteristics. One option is to adopt a probabilistic interpretation for all calculated elastic acceleration response spectra on the ground surface with predefined exceedance probability to define the elastic acceleration spectra to be used for microzonation and for vulnerability assessment.

The input acceleration time histories are scaled for each cell with respect to the peak accelerations obtained from earthquake hazard study since this approach was observed to be practical and gave consistent results as shown by Durukal et al. (2006) and Ansal et al. (2006b). For the case of the Bolu City microzonation study 28 scaled acceleration time histories were used as input motion for site response analyses by Shake91 (Idriss and Sun, 1992) and the average of the acceleration response spectra on the ground surface were determined to obtain the necessary parameters for microzonation as well as for the vulnerability assessment of the building stock.

3 Local Site Conditions

In the second stage of the microzonation studies, the investigated regions are divided into cells by a grid system depending on the availability of geological, geophysical and geotechnical data. During recent studies, grid systems with 250 m × 250 m cells are adopted and recommended to define the site conditions in terms of representative soil profiles for each cell by investigating and evaluating the existing data.

Two issues are important in the determination of local site conditions: The first one is the soil classification for each layer encountered within the soil profile based on the results of laboratory index tests (grain size distribution and Atterberg limits) performed on samples obtained from borings and the second one is the depth of engineering bedrock, which can be defined as the layer with shear wave velocity, $V_S \geq 750$ m/s. In order to determine the depth of engineering bedrock, deep boreholes and/or geophysical insitu tests should be used. This information should have been compiled to perform site classification based on various earthquake codes and to carry out site response analysis.

In the case of limited experimental and numeric data, an effort should be made to be on the safe side by selecting softer and weaker site classes. However, this conservatism does not always guarantee to be on the safe side. Site classifications, site response, liquefaction, and landslide analysis may not be realistic if conducted based on incomplete and unsuitable data. Therefore it is crucial that microzonation studies should be carried out based on detailed geological, geotechnical and geophysical site investigations supported by soil borings and insitu tests performed in each cell of the grid system.

Site conditions can be classified using the representative soil profiles selected for each cell where one or more borehole data is available based on the detailed assessment of the available geological and geotechnical data (soil stratification, layer thicknesses, soil types, engineering bedrock depth, and variation of shear wave velocity with depth) according to national (TEC, 2007), European (CEN, 2006) and American (NEHRP, 2003) earthquake codes.

The soil classification based on different earthquake codes is a Grade 1 type of microzonation with respect to ground shaking intensity (ISSMGE/TC4, 1999). This zonation maps based on site classifications are very similar to the zonation maps developed based on geological formations. Zonation maps either based on site classification or geologic formations are very rough due to the fact that both site classification in the earthquake codes and geological formations are defined within relatively large ranges and in addition only involves one part of the microzonation problem by neglecting earthquake source characteristics. If these two microzonation approaches are compared, the microzonation with respect to earthquake codes is generally more preferable than the microzonation with respect to geologic formations, in view of the fact that corresponding to each site classification in the earthquake codes, structural design criteria is specified based on the past experience and the results of parametric and observational studies. Such justifications can not be given for microzonation with respect to geological formations and that is why microzonation maps only with respect to geological formations are not sufficient anymore. One positive aspect of microzonation maps with respect to site classification is that they can be used as guidelines for the future site specific studies, since they contain general information on site classes.

As an example given in Fig. 1, the site classification at each cell was established according to Turkish Earthquake Code, TEC (2007) for Bolu City. Turkish Earthquake Code site classification is a two stage classification; the first stage requires the definition on soil types for each layer where soil types are defined with respect to their shear strength and density from A to E, and the second stage is based on layer thickness of different soil types and classification ranges from Z1 (rock or hard/very dense shallow deposits) to Z4 (soft and deep deposits).

For site classification, all available data from individual geotechnical investigation studies in Bolu city was compiled. Yet, the number of cells having at least one borehole was only 211 among the total number of cells (377) for the whole study area. Representative soil profiles for each cell where one or more borehole data was available were selected by considering the most unfavorable profile to be on the safe side. As can be seen from Fig. 1, there is no settlement in the cells lacking any

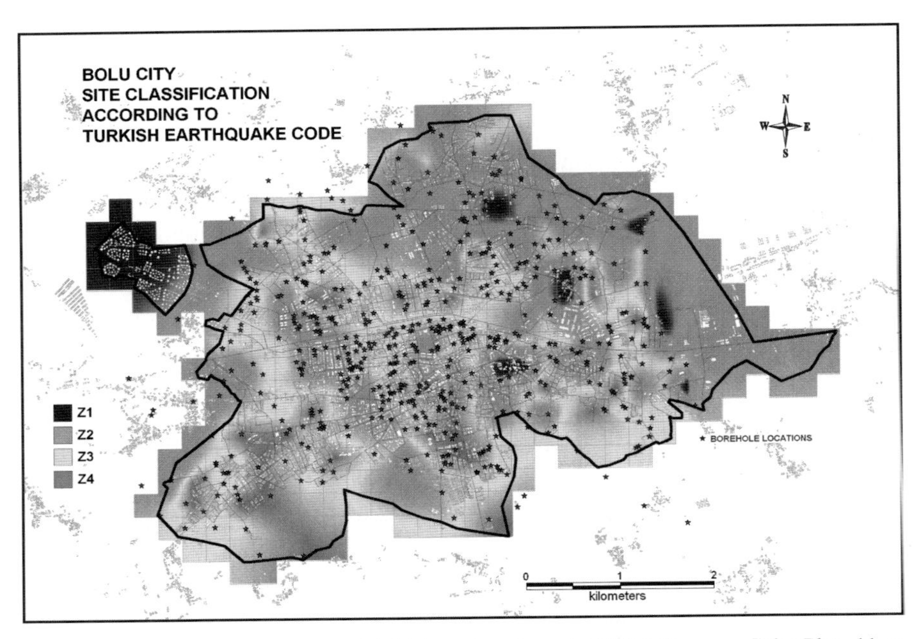

Fig. 1 Microzonation with respect to Turkish earthquake code (2007) (See also Color Plate 11 on page 213)

available borehole information. For this reason, the cells with no available borehole information were not evaluated and interpolations between neighboring boreholes were performed using the GIS mapping software package in establishing the zonation map. As shown again in Fig. 1, instead of using rigid boundaries, soft transitions were preferred to map the variation of microzonation parameters. The first reason of this preference is the difficulties and ambiguities in determination of contours as fixed boundaries. The second reason is to provide some flexibility to authorities that will be conducting urban planning based on these microzonation maps.

To perform site response analysis, soil stratification, layer thicknesses, soil types, shear wave velocity or small strain shear modulus (G_{MAX}) for each layer and relationships of shear modulus and damping as a function of shear strain for each soil type should be provided (Vucetic and Dobry, 1991, Okur and Ansal, 2007). The variations of shear wave velocities with depth can be determined from in-situ seismic wave velocity measurements, if available, or from SPT blow counts using empirical relationships proposed in the literature. During microzonation studies carried out for Bolu City, the variation of shear wave velocities with depth were determined from SPT blow counts using the empirical relationship proposed by Iyisan (1996);

$$V_s = 51.5\,N^{0.516} \leq 500\,\mathrm{m/s} \tag{1}$$

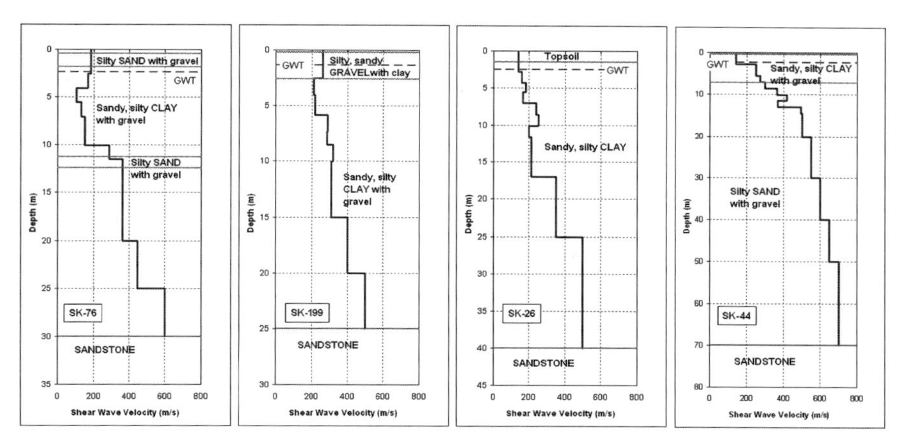

Fig. 2 Typical soil profiles with the variation of shear wave velocity with depth

in terms of uncorrected standard penetration blow counts, N. Shear wave velocity profiles were established down to the engineering bedrock with estimated shear wave velocity of $V_S = 750\,m/s$ and representative soil profiles were modeled. Some typical shear wave velocity profiles form Bolu City determined in this manner are shown in Fig. 2.

Both the depth of engineering bedrock and the variation of shear wave velocities with depth can also be determined from in-situ geophysical wave velocity measurements. These methods should be applied as much as possible and the calculated shear wave velocity profiles should be compared with respect to measured shear wave velocity data obtained by in-situ seismic wave velocity measurements and calculations should be modified when necessary.

Research on strong ground motion records obtained from vertical arrays has shown that the earthquake characteristics on the ground surface are affected significantly by the soil layers at the top 30 m of the soil profile, which indicates that a parameter such as weighted average of the shear wave velocities for the top 30 m (defined as average shear wave velocity) could be very useful in evaluating the effects of site conditions (Borcherdt, 1994). In other words, soil amplification can be estimated approximately by using equivalent shear wave velocity. Therefore, the site classification definitions in several earthquake codes (e.g. NEHRP, 2003; CEN EC8, 2006) are based on the average shear wave velocity for the top 30 m and as such this weighted average (equivalent) of the shear wave velocities for the top 30 m can also be used as a microzonation parameter.

The development of microzonation maps involves the division of the area into three zones (as A, B, and C) based on relative values of the parameters considered (Ansal et al., 2004b). The site characterizations, as well as all the analyses performed, require various approximations and assumptions, thus it is recommended not to present the numerical values for any parameter since their relative values are more important then their absolute values. In all cases, the variations of the calculated parameters are considered separately and their frequency distributions

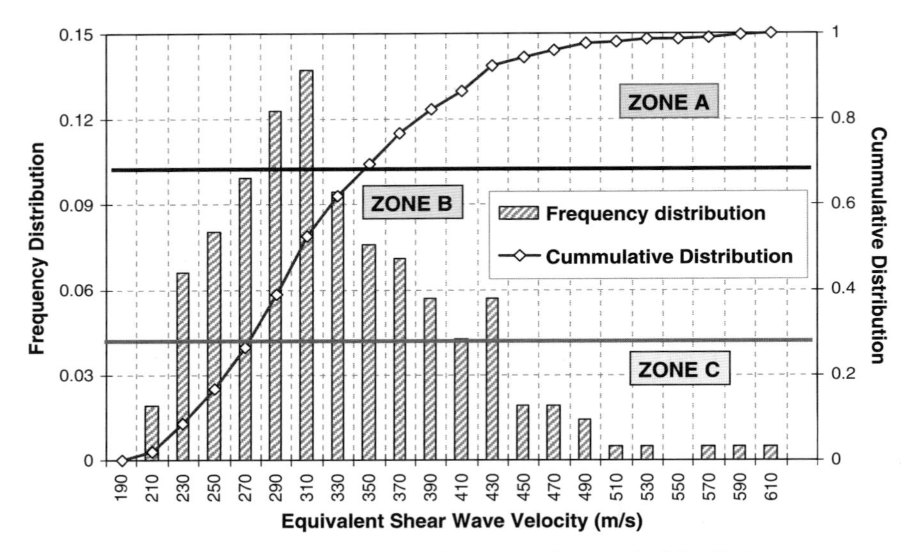

Fig. 3 Relative zonation approach adopted with respect to the statistical distribution

are determined to calculate the 33% and 67% percentiles to define the boundaries between the three zones as shown in Fig. 3. The zone A shows the most favorable 33% percentile (i.e. low spectral accelerations), zone B the medium 34% percentile and zone C shows the most unsuitable 33% percentile (i.e. high spectral accelerations).

Microzonation with respect to average (equivalent) shear wave velocity is more preferable compared to microzonation based on site classification (Fig. 1) because relative mapping approach by dividing the area into three zones is only possible for the case of microzonation with respect to average shear wave velocity. The advantage of relative zonation is to enable city planners and city authorities of various professions to understand the microzonation logic without much technical knowledge.

In an urban area, the earthquake hazard level may not alway be very high; even if it is high, with an appropriate design it is possible to construct any type of building anywhere as desired. The important thing in urban planning is the identification of areas of relative risk to enable the structural designer and builder to anticipate related problems to design and build accordingly.

A zonation map based on average shear wave velocity is given in Fig. 4 for the case of the Bolu City. However, these maps also reflect only one aspect of the microzonation methodology neglecting earthquake characteristics. For the case of Bolu City, during the preparation of microzonation maps with respect to average shear wave velocity, only cells with measured or calculated data were utilised and for the cells with no data interpolation between neighboring cells was performed using the GIS mapping as shown in Fig. 4.

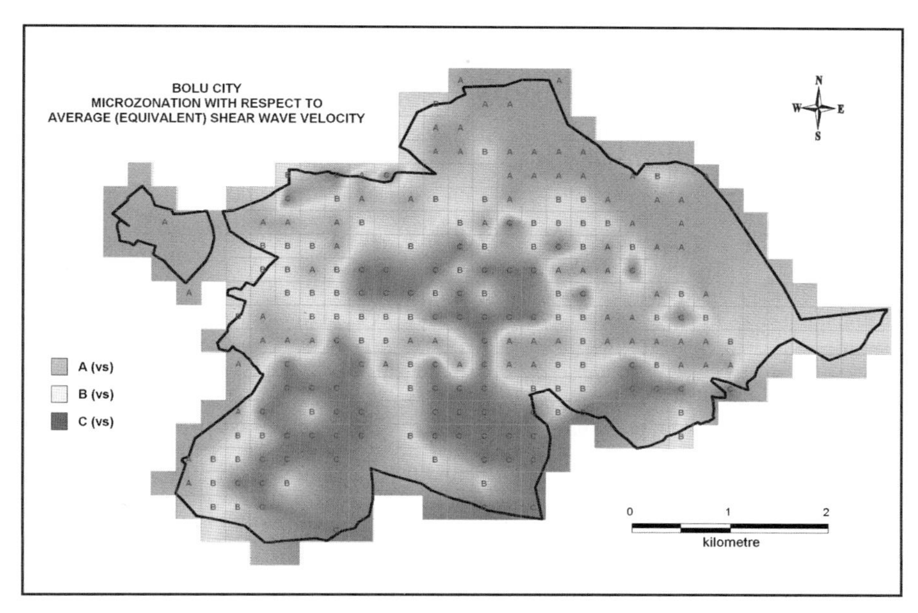

Fig. 4 Microzonation with respect to average shear wave velocity (V_{S30}) (See also Color Plate 12 on page 213)

4 Site Response Analysis

For all soil layers in a soil profile; soil type, thickness, total unit weight, shear wave velocity, and G/G_{MAX} and damping relationships need to be provided as input to be used in site response analysis. For site response analysis, selection of strain dependent shear modulus and damping ratio relationships appropriate for that particular soil type affects results as much as soil stratification and thickness of the layers.

Earthquake characteristics (peak ground accelerations and elastic acceleration response spectra) on the ground surface are determined by conducting one dimensional site response analysis for soil profiles of each cell for all selected input motion records. All previously recorded strong ground motion time histories compatible with the earthquake hazard assessment in terms of possible magnitude; distance and fault mechanism selected as the input rock outcrop motion are scaled for each cell with respect to the peak accelerations obtained from earthquake hazard study. The average acceleration response spectra calculated on the ground surface using all scaled input acceleration time histories as input in site response analyses can be used as the response spectra corresponding to earthquake hazard scenario spectrum for each cell.

Studies on microzonation with respect to peak spectral accelerations using two different sets of three real (compatible with the earthquake hazard for Zeytinburnu) that were scaled with respect to the identical peak accelerations and one set of three simulated (compatible with the time dependent earthquake hazard spectra) acceleration time histories reveal that independent than the scenario selected, acceleration time histories used in site response analysis, in other words source characteristics are very important. Acceleration time histories recorded during same or different earthquakes on different site conditions may be very different and can introduce significant variability in engineering applications. One approach is to adopt a probabilistic interpretation of the calculated elastic acceleration response spectra from all site response analyses using as much as possible large number of real input acceleration records obtained on compatible tectonic, seismic and site conditions (Ansal and Tönük, 2007b). This approach has the advantage of defining the hazard level in accordance with the purpose of the microzonation.

For Bolu City microzonation study, 28 strong ground motion records compatible with local seismic hazard in terms of magnitude, distance and fault mechanism were selected to be used in site response analysis (PEER). The selected records scaled with respect to peak accelerations for each soil profile obtained from earthquake

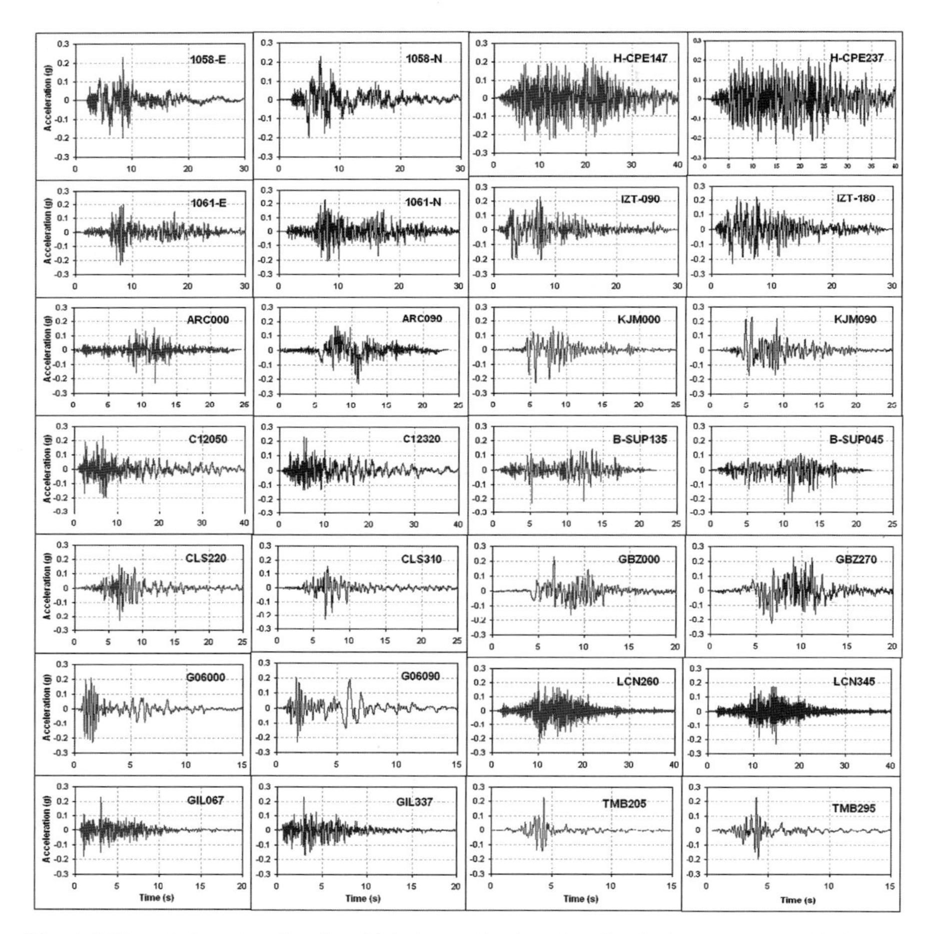

Fig. 5 PGA scaled acceleration time histories used as input motion in site response analysis

hazard study as proposed by Ansal et al. (2006b) and Durukal et al. (2006) are shown in Fig. 5. These 28 scaled acceleration time histories were used as input motion for site response analyses for each cell and the average of the acceleration response spectra on the ground surface were determined to obtain the necessary microzonation parameters.

5 Seismic Microzonation with Respect to Ground Shaking

During the last decade, the increase in the analytical, in-situ and laboratory investigation capabilities for assessment of local site conditions and site response analysis has affected significantly the development of microzonation techniques. Studies on the observed damage distribution and the strong ground motion records obtained during recent earthquakes (Gazetas et al., 1990; Faccioli, 1991; Ansal et al., 1993; Bard, 1994; Chavez-Garcia et al., 1996; Chin-Hsiung et al., 1998; Kawase, 1998; Athanasopoulus et al., 1999; Hartzell et al., 2001) have shown that earthquake source mechanisms and the regional site conditions are important controlling factors.

The intention in assessing the ground shaking intensity is to estimate the effects of local site conditions. Thus, it would be logical to base this decision on all the available results from site identifications based on both equivalent shear wave velocity and site response analysis conducted for each cell. Among these two approaches, as the first microzonation parameter, the peak spectral accelerations (at $T = 0.2$ s) calculated from Borcherdt (1994) using equivalent shear wave velocities were adopted and zonation map was produced in accordance with the relative mapping in terms of three zones. For Bolu City case, however, since the difference between peak spectral accelerations (at $T = 0.2$ s) calculated from Borcherdt (1994) corresponding 33% and 67% percentiles of the distribution was smaller than 20%, the area was divided into two zones using 50% percentile (median) value of 0.598 g as recommended by Studer and Ansal (2004). In Fig. 6, A_{SA} borch shows the most favorable regions with lower 50% percentile where spectral accelerations are less than 0.598 g and C_{SA} borch shows the most unsuitable regions with higher 50% percentile with respect to peak spectral accelerations where the spectral accelerations are higher than 0.598 g.

For microzonation with respect to ground shaking intensity, the second microzonation parameter adopted is the average spectral accelerations calculated between the 0.1 and 1 s periods using the average acceleration spectra determined from the results of the all site response analyses conducted for each cell. The range of average spectral accelerations computed for the period interval of 0.1–1.0 s was between 0.354 and 0.797 g for the Bolu City case and again as in the first microzonation parameter (peak spectral accelerations calculated from based on average shear wave velocity), since the difference between 33% and 67% percentiles was smaller than 20%, the area was divided into two zones using 50% percentile (median) value of 0.575 g with respect to average spectral accelerations. In Fig. 7, A_{SA} ave shows the most favorable regions with lower 50% percentile and C_{SA} ave shows the most unsuitable regions with higher 50% percentile with respect to average spectral accelerations.

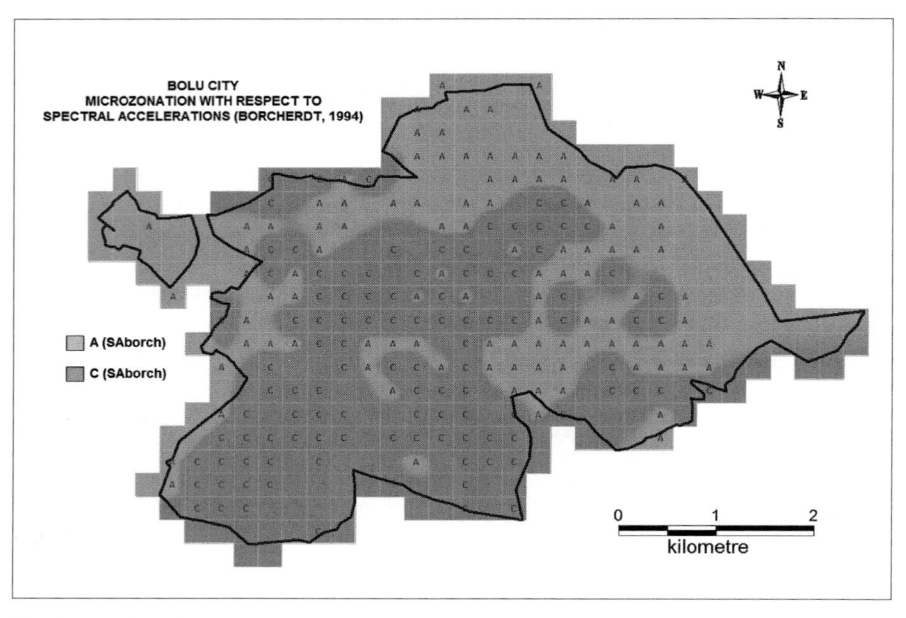

Fig. 6 Zonation with respect to spectral accelerations calculated using Borcherdt (1994) procedure for Bolu city (See also Color Plate 13 on page 213)

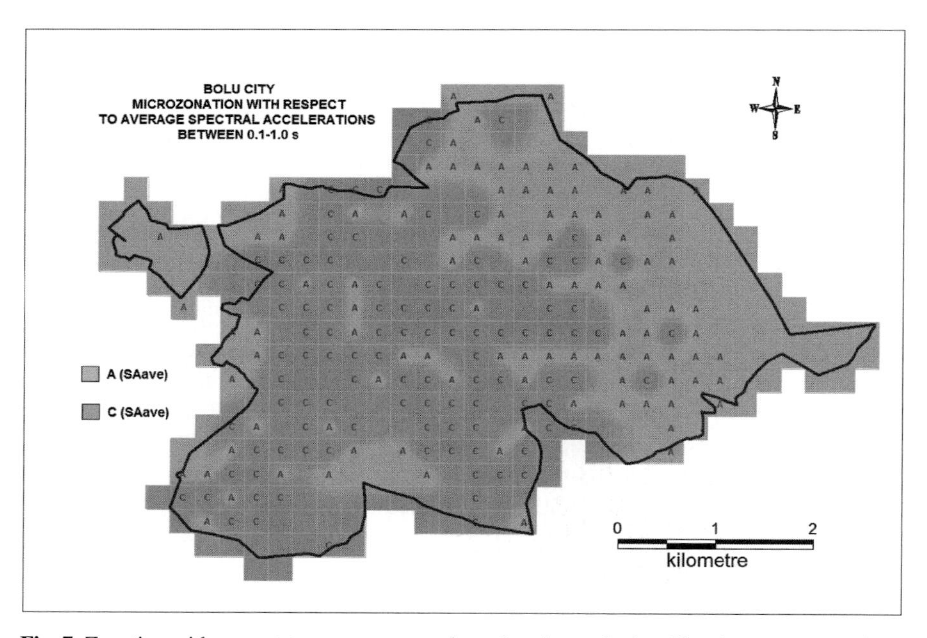

Fig. 7 Zonation with respect to average spectral accelerations calculated by site response analysis for Bolu city (See also Color Plate 14 on page 213)

As can be seen from these maps (Figs. 6 and 7), there are similarities and differences between the average spectral accelerations obtained by site response analyses with the spectral accelerations calculated using Borcherdt (1994) based on equivalent shear wave velocity. The site response analysis, independent from the software used, would sometimes give unrealistically high spectral amplifications or very low peak ground acceleration values depending on the thickness of the deposit, estimated initial shear moduli, and also on the characteristics of the input acceleration time histories. On the other hand, even though they are more empirical, the spectral accelerations calculated using equivalent shear wave velocities tend to give more consistent values that appear to be more realistic when compared with the selected soil profiles. Thus, the final ground shaking microzonation map is produced by the superposition of these two maps taking into consideration both approaches (Ansal et al., 2004b, Studer and Ansal, 2004).

In the final ground shaking microzonation map where A_{GS} shows the areas with lower ground shaking and C_{GS} shows the areas with higher ground shaking potential. The zone A_{GS} corresponds to overlapping zones of (A_{SA} ave and A_{SA} borch). The zone B_{GS} corresponds to overlapping zones of (A_{SA} ave and C_{SA} borch) or (C_{SA} ave and A_{SA} borch). The zone C_{GS} corresponds to overlapping zones of (C_{SA} ave and C_{SA} borch). The assessment for each cell is performed numerically by adopting the above criteria to determine the three zones and to perform the mapping using the new data. At this final stage, it would be recommended to compare the ground shaking zonation map with the surface geology map as shown in Fig. 8 to verify the reliability of the proposed ground shaking intensity microzonation map.

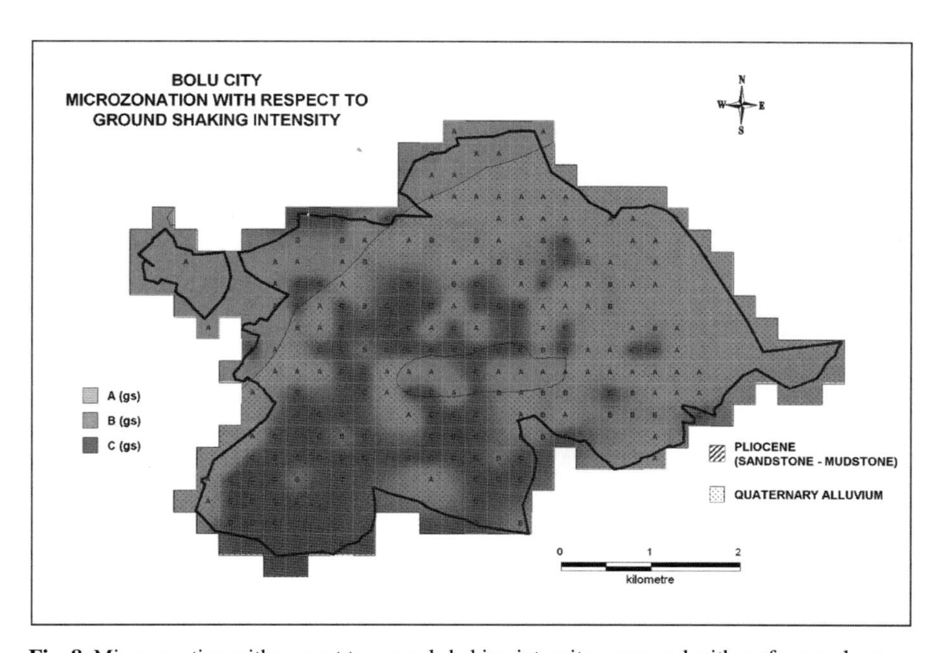

Fig. 8 Microzonation with respect to ground shaking intensity compared with surface geology

Seismic microzonation can be defined as the process for estimating the response of soil layers under earthquake excitations and the variation of earthquake ground motion characteristics on the ground surface (ISSMGE/TC4, 1999). Thus, one of the main targets is an evaluation based on ground shaking intensity (Marcellini et al., 1995; Lachet et al., 1996; Fäh et al., 1997). However, this does not involve structural damage estimation. Structural damage occurred during an earthquake is a complex function of three coupling factors: source characteristics, local site conditions and structural features including foundations. Therefore, in order to model damage distribution, these coupling factors should be evaluated all together. On the other hand, seismic zonation reflects only two of the three factors and because of that should not be used alone for damage estimation.

6 Seismic Microzonation with Respect to Spectral Accelerations

Another purpose of microzonation is to provide an input for earthquake damage scenarios. Acceleration response spectrum is mostly used parameter for the assessment of the vulnerability of the building stock for different hazard (performance) levels. The earthquake characteristics used in the assessment of the structural vulnerability may be calculated based on the conventional NEHRP (2003) procedure considering the microzonation map obtained in terms of NEHRP site classification. The spectral accelerations may also be calculated by site response analyses. The average acceleration response spectra obtained for each cell from site response analyses are evaluated for determining the spectral accelerations for the short period (Ss) corresponding to $T = 0.2$ s and for the long period (S1) corresponding to $T = 1$ s. An optimization algorithm is used to determine the best fit envelope to the calculated average acceleration response spectra (Ansal et al., 2005a). All the requirements of the NEHRP spectra are adopted where the two independent variables in the developed optimization algorithm are (Ss) and (S1). Some examples of the best fit envelopes obtained by this approach with respect to average acceleration response spectra are shown in Fig. 9 together with the acceleration response spectra calculated using the selected PGA scaled input acceleration time histories.

A parametric study was conducted to observe the effect of averaging and adopting a probabilistic approach (10% exceedance) by assuming that the calculated

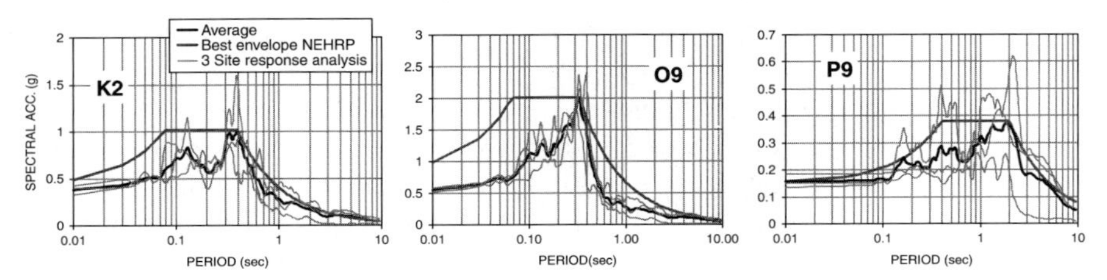

Fig. 9 Some typical best envelop NEHRP spectra fitted to average acceleration response spectra and acceleration response spectra calculated by site response analysis

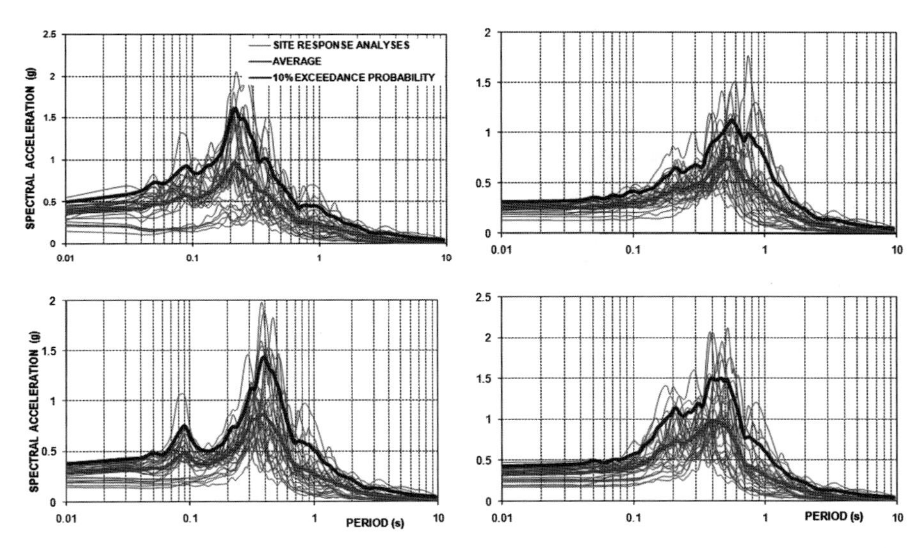

Fig. 10 Examples of 10% probability of exceedance elastic acceleration response spectra, average and 28 acceleration response spectra calculated by site response analysis for different cells

elastic acceleration spectra can be modeled to have normal distribution for the period range considered. A few examples of the average elastic acceleration response spectra are shown in Fig. 10, this time in comparison with all the acceleration response spectra calculated for 28 PGA scaled input acceleration time histories and 10% exceedance probability acceleration spectra calculated assuming that spectral accelerations for each period have normal distribution. In Fig. 11, zonation maps with respect to peak spectral acceleration obtained from the average spectra of all site response analyses and spectra corresponding to 10% exceedance probability are compared for Zeytinburnu case study (Ansal and Tönük, 2007b).

Microzonation maps with respect to peak spectral accelerations, varying between 0.6 and 3.0 g, were plotted in increments of 0.2 g. As can be observed in Fig. 11, there are significant differences between these two zonation maps arising from probabilistic evaluation of the median and 10% exceedance levels. However, when more conservatism is preferred, the probabilistic approach adopted in this scheme with the same hazard level (10% exceedance probability) that was used to calculate the earthquake hazard on the rock outcrop for site response analyses may be used.

The comparison between the spectral accelerations obtained from site response analyses using the best envelope fitting procedure with the values obtained by the NEHRP formulation indicate that the values obtained by site response analyses shows much larger range. The difference in the data range is much more significant in the case of short period spectral accelerations. This may be an indication of more accurate determination of site effects and is partly due to the relatively large shear wave velocity ranges in the NEHRP site classes.

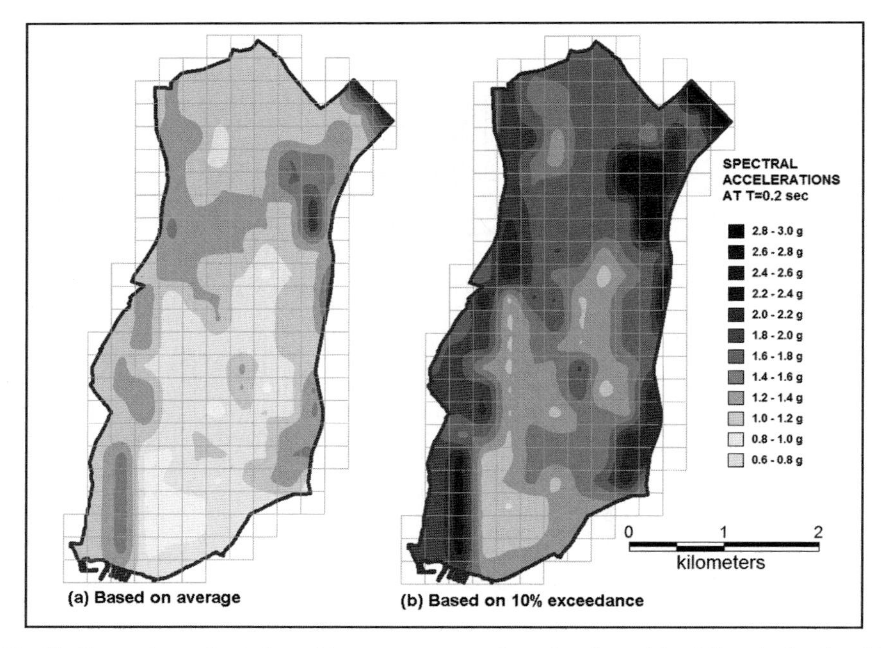

Fig. 11 Comparison of zonation with respect to spectral acceleration at $T = 0.2$ s calculated as (**a**) the average off all site response analyses and (**b**) corresponding to 10% exceedance probability for Zeytinburnu

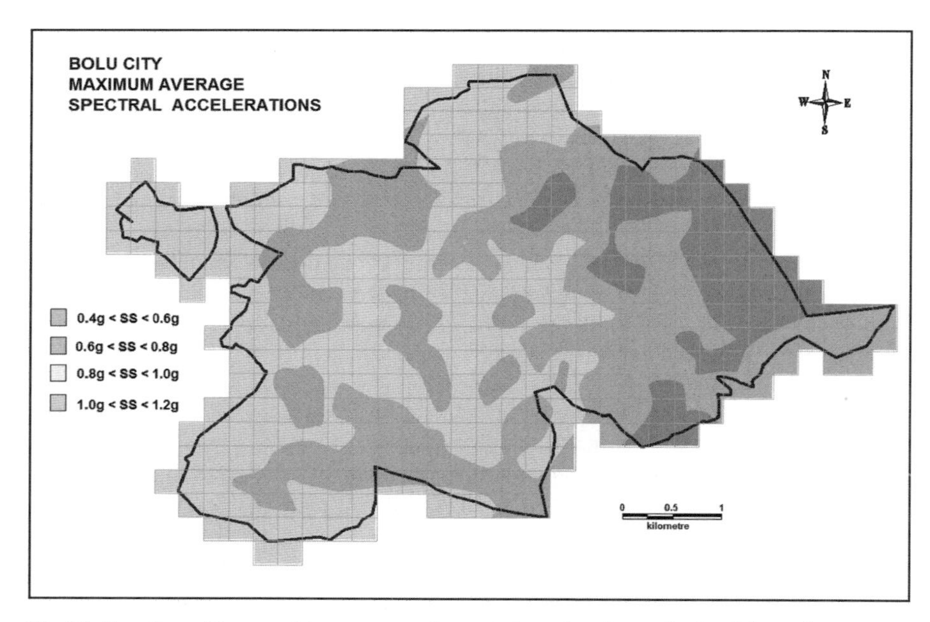

Fig. 12 Zonation with respect to average peak spectral accelerations calculated from site response analysis for the Bolu city

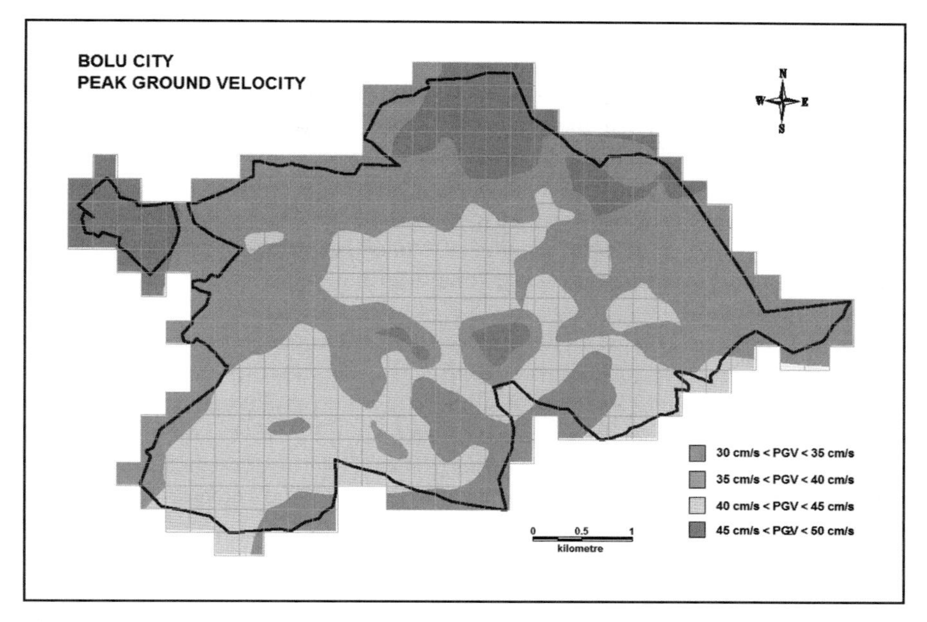

Fig. 13 Zonation with respect to average peak ground velocity based on site response analysis for the Bolu city

The variations of average peak spectral accelerations and average peak ground velocities calculated for 28 PGA scaled input acceleration time histories on the ground surface within Bolu city are shown in Figs. 12 and 13, respectively. These zonation maps show that the variation of spectral accelerations used for the assessment of the vulnerability of the building stock may be significantly different than the variation of peak ground velocities used for the assessment of the vulnerability of the lifeline systems.

7 Conclusions

The procedure proposed for conducting a microzonation study with respect to earthquake hazard scenario is based on a grid system preferable with cells of 250 m × 250 m. In the first stage, regional seismic hazard analyses need to be conducted to estimate earthquake characteristics on rock outcrop for each cell by a probabilistic approach at the regional level to determine the earthquake characteristics for 475 years return period corresponding to 10% exceedance in 50 years. For each cell, representative soil profiles down to engineering bedrock depth need to be determined based on detailed assessment of available geotechnical data. Representative soil profiles of cells where one or more borehole data were available were created by considering the most suitable boring, and for the cells with no borehole information available representative soil profiles were either selected from the neighboring cells by utilizing the available data or left as blank cells.

Microzonation with respect to ground shaking intensity was based on two parameters: (1) average spectral accelerations calculated between the 0.1 and 1 s periods using the average acceleration spectrum calculated for each cell from the results of the site response analysis conducted for each cell, (2) the peak spectral accelerations calculated from Borcherdt (1994) using equivalent shear wave velocities. The microzonation with respect to ground shaking intensity is produced by the superposition of these two maps with respect to three regions where zone A_{GS} shows the areas with very low ground shaking intensity, zone B_{GS} shows the areas with low to medium ground shaking intensity, and zone C_{GS} shows the areas with high ground shaking intensity.

For the assessment of the vulnerability of the building stock, the average acceleration response spectra obtained for each cell from site response analyses were evaluated for determining the spectral accelerations for the short period corresponding to $T = 0.2$ s and for the long period corresponding to $T = 1$ s. An approach was adopted to determine the best fitting NEHRP envelope to the calculated average acceleration response spectra. The variation of peak ground velocities are used for the assessment of the vulnerability of the lifeline systems. It appears possible to assume that it would be more reliable to perform site response analyses to determine spectral accelerations and peak ground velocity on the ground surface to be used as input to the vulnerability studies depending on the quality and accuracy of the building and lifeline stock information.

References

Ansal, A., Akinci, A., Cultrera, G., Erdik, M., Pessina, V., Tönük, G., and Ameri, G. (2009) "Loss Estimation in Istanbul Based on Deterministic Earthquake Scenarios of the Marmara Sea Region (Turkey)" Soil Dynamics and Earthquake Engineering, 29(4), 699–709.

Ansal, A., Kurtuluş, A., and Tönük, G. (2008) "Damage To Water And Sewage Pipelines In Adapazari During 1999 Kocaeli, Turkey Earthquake", Proceedings of 6th International Conference on Case histories in Geotechnical Engineering, Arlington, Virginia, ABD, Paper No:3.17.

Ansal, A., Tönük, G., and Kurtuluş, A. (2007c) "Microzonation with Respect to Ground Shaking Intensity and Seismic Hazard Scenarios", Invited Lecture, Sixth National Conference on Earthquake Engineering, Istanbul, Turkey, V3:133–151.(in Turkish).

Ansal, A., Tönük, G., and Bayraklı, Y. (2007b), "Microzonation with Respect to Ground Shaking Intensity Based on 1D Site Response Analysis", XIV European Conference on Soil Mechanics and Geotechnical Engineering, Madrid, Spain.

Ansal, A., Kurtuluş, A., and Tönük, G. (2007a), "Earthquake Damage Scenario Software for Urban Areas", Keynote Lecture, Int. Con. on Computational Methods in Structural Dynamics and Earthquake Engineering, M. Papadrakakis, D.C. Charmpis, N.D. Lagaros, Y. Tsompanakis (eds.), Rethymno, Crete, Greece.

Ansal, A. and Tönük, G. (2007b), "Source and Site Effects for Microzonation", Theme Lecture, 4th International Conference on Earthquake Geotechnical Engineering, Earthquake Geotechnical Engineering, K. Pitilakis (ed.), Chapter 4, 73–92, Springer.

Ansal, A. and Tönük, G. (2007a) "Ground Motion Parameters for Loss Estimation", Keynote Lecture, Fourth International Conference on Urban Earthquake Engineering, Tokyo Institute of Technology, Tokyo, Japan, 7–14.

Ansal, A., Durukal, E., and Tönük, G. (2006b), "Selection and Scaling of Real Acceleration Time Histories for Site Response Analyses", Proceedings of ETC12 Workshop, Athens, Greece.

Ansal, A., Tönük, G., Demircioğlu, M., Bayraklı, Y., Şeşetyan, K., Erdik, M. (2006a) "Ground Motion Parameters for Vulnerability Assessment", Proceedings of the First European Conference on Earthquake Engineering and Seismology, Geneva, Switzerland, Paper Number: 1790.

Ansal, A., Erdik, M., Eyidoğan, H., Özaydin, K., Yildirim, M., Siyahi, B. (2005b), "Special Technical Requirements for Microzonation Maps", Istanbul Metropolitan Municipality, (in Turkish).

Ansal, A., Özaydın, K., Erdik, M., Yıldırım, Y., Kılıç, H., Adatepe, Ş., Özener, P.T., Tonaroğlu, M., Şeşetyan, K., Demircioğlu, M. (2005a), "Seismic Microzonation for Urban Planning And Vulnerability Assessment", Proceedings of the International Symposium of Earthquake Engineering (ISEE2005), Awaji Island, Kobe, Japan.

Ansal, A., Erdik, M., Studer, J., Springman, S., Laue, J., Buchheister, J., Giardini, D., Faeh, D., and Koksal, D. (2004c), "Seismic Microzonation For Earthquake Risk Mitigation In Turkey" Proceedings of the 13th World Conference of Earthquake Engineering, Vancouver, Canada, CD paper No.1428.

Ansal, A., Laue, J., Buchheister, J., Erdik, M., Springman, S.M., Studer, J., and Koksal, D. (2004b), Site Characterization and Site Amplification for a Seismic Microzonation Study in Turkey. Proceedings of 11th International Conference on Soil Dynamics and Earthquake Engineering and 3rd Earthquake Geotechnical Engineering, San Francisco, USA.

Ansal, A., Biro, Y., Erken, A., and Gulerce, U. (2004a), "Seismic Microzonation: A Case Study", Chapter 8, Recent Advances in Earthquake Geotechnical Engineering and Microzonation, Kluwer Academic Publications.

Ansal, A. (2002), "Seismic Microzonation Methodology", Proceedings of 12th European Conference on Earthquake Engineering, Paper No.830, London, UK.

Ansal, A.M., Şengezer, B.S., İyisan, R., and Gençoğlu, S. (1993), "The Damage Distribution in March 13, 1992 Earthquake and Effects of Geotechnical Factors", Soil Dynamics and Geotechnical Earthquake Engineering, Ed.P.Seco e Pinto, Balkema, 413–434.

Athanasopoulus, G.A., Pelekis, P.C., and Leonidou, E.A. (1999), "Effects of Surface Topography on Seismic Ground Response in the Egion (Greece), 15 June 1995 Earthquake", Soil Dynamics and Earthquake Engineering, 18, 135–149.

Bard, P.-Y. (1994), "Effects of Surface Geology on Ground Motion: Recent Results and Remaining Issues", Proceedings of the 10th ECEE, Vienna, Austria, 1, 305–323.

Borcherdt, R.D. (1994), Estimates of Site Dependent Response Spectra for Design (Methodology and Justification), Earthquake Spectra, 10(4), 617–654.

CEN (2006), Eurocode EC8 – Design of structures for earthquake resistance, European Standard, European Committee for Standardisation, Brussels.

Chavez-Garcia, F.J., Cuenca, J., and Sanchez-Sesma, F.J. (1996), "Site Effects in Mexico City Urban Zone, A Complementary Study", Soil Dynamics and Earthquake Engineering, 15, 141–146.

Chin-Hsiung, L., Jeng-Yaw, H., and Tzay-Chyn, S. (1998), "Observed Variation of Earthquake Motion across a Basin-Taipei City", Earthquake Spectra, 14(1), 115–134.

Durukal, E., Ansal, A., and Tönük, G. (2006), "Effect of Ground Motion Scaling and Uncertainties in Site Characterisation on Site Response Analyses" Proceedings of Eighth U.S. National Conference on Earthquake Engineering, San Francisco, California.

Erdik, M., Demircioglu, M., Sesetyan, K., Durukal, E., and Siyahi, B. (2004), "Earthquake Hazard in Marmara Region", Soil Dynamics and Earthquake Engineering, 24, 605–631.

Faccioli, E. (1991), "Seismic Amplification in the Presence of Geological and Topographic Irregularities", Proceedings of the 2nd International Conference on Recent Advances in Geotechnical Earthquake Engineering, St. Louis, Missouri, State-of-the-Art paper, 1779–1797.

Fäh, D., Rüttener, E., Noack, T., and Kruspan, P. (1997), "Microzonation of the City of Basel", Journal of Seismology, 1, 87–102.
Gazetas, G., Dakoulas, P., and Papageorgiou, A. (1990), "Local Soil and Source-Mechanism Effects in the 1986 Kalamata (Greece) Earthquake" Earthquake Engineering and Structural Dynamics, 19, 431–453.
Hartzell, S., Carandr, D., and Williams, R.A. (2001), "Site Response, Shallow Shear-Waand Velocity and Damage in Los Gatos, California, from the 1989 Loma Prieta Earthquake", BSSA, 91(3), 468–478.
Idriss, I.M. and Sun, J.I. (1992), "Shake91, A Computer Program for Conducting Equivalent Linear Seismic Response Analysis of Horizontally Layered Soil Deposits Modified based on the original SHAKE program Published in December 1972 by Schnabel, Lysmer and Seed".
ISSMGE/TC4 (1999), Manual for Zonation on Seismic Geotechnical Hazards, Tech. Com. For Earthquake Geotechnical Engineering TC4, ISSMGE, Japanese Society of Soil Mechanics..
Iyisan, R. (1996) "Correlations between shear wave velocity and in-situ penetration test results", Technical Journal of Turkish Chamber of Civil Engineers 7(2), 1187–1199 (in Turkish).
Kawase, H. (1998), "The Cause of the Damage Belt in Kobe: 'The Basin Edge Effect', Constructive Interference of the Direct S-Wave with the Basin-induced Diffracted/Rayleigh Waves", Seismological Research Letters, (67), 25–34.
Kılıç, H., Özener, P.T., Ansal, A., Yıldırım, M., Özaydın, K., Adatepe S (2006), Microzonation of Zeytinburnu Region with respect to Soil Amplification: A Case Study. Journal of Engineering Geology, 86, 238–255.
Lachet, C., Hatzfeld, D., Bard, P.Y., Theodulidis, N., Papaioannou, C. & Savvaidis, A. (1996), "Site Effects and Microzonation in the City of Thessaloniki-Comparison of Different Approaches", BSSA, 86(6), 1692–1703..
Marcellini, A., Bard, P.Y., Iannaccone, G., Meneroud, J.P., Mouroux, P., Romeo, R.W., Silandstri, F., Duval, A.M., Martin, C., and Tento, A. (1995), "The Benevento Seismic Risk Project. II-The microzonation", Proc. 5th International Conference on Seismic Zonation, Nice, France, 1, 810–817..
NEHRP (2003), Recommended Provisions for New Buildings and Other Structures, FEMA-450, prepared by the Building Seismic Safety Council for the Federal Emergency Management Agency, Washington, DC.
Okur, D.V. and Ansal, A. (2007), "Stiffness Degradation of Natural Fine Grained Soils During Cyclic Loading", Soil Dynamics and Earthquake Engineering, 27(9), 843–854.
PEER Strong Motion Data Bank, http://peer.berkeley.edu.
Spence, R. (editör) (2007), Earthquake Disaster Scenario Prediction and Loss Modelling for Urban Areas, EU LessLoss Report No.7, EU FP7 Project.
Studer, J. and Ansal, A. (2004), "Seismic Microzonation for Municipalities, Manual." Research Report for Republic of Turkey, Ministry of Public Works and Settlement, General Directorate of Disaster Affairs, World Institute for Disaster Risk Management, Inc.
Turkish Earthquake Code (TEC) (2007), Specification for Buildings to be Built in Earthquake Areas. Ministry of Public Works and Settlement, Government of Republic Turkey.
Vucetic, M. and Dobry, R. (1991), "Effect of Soil Plasticity on Cyclic Response", Journal of Geotechnical Engineering, ASCE, 117(1), 89–107.

Performance of Structures During the 2004 Indian Ocean Tsunami and Tsunami Induced Forces for Structural Design

Murat Saatçioğlu

Abstract Observations from a reconnaissance visit to the 2004 Indian Ocean Tsunami disaster area are presented from structural performance perspective. Behaviour of engineered and non-engineered buildings under tsunami-induced hydrodynamic and debris impact forces is discussed. It is shown that engineered reinforced concrete buildings generally survived the tsunami without much damage. Non-engineered concrete frame and confined masonry buildings suffered different degrees of structural and non-structural damage, depending on the topography and their proximity to the shoreline. Low-rise timber construction exhibited little resistance to tsunami loads, resulting in massive destruction of large residential areas. Recommendations made by existing building codes for tsunami force computations are reviewed. A case study is presented for a 10-storey reinforced concrete building located along the Pacific coast of Canada in terms of tsunami and seismic design force levels. It is demonstrated that local structural damage can be induced by tsunami forces on vertical elements designed to perform within the inelastic range of deformations under earthquake-induced forces.

1 Introduction

Tsunami-induced forces are often neglected in structural design practice. This may be attributed to long return periods of large magnitude tsunamis in populated and built environments. Indeed, the return period of major tsunamis may be in excess of 500 years. However, the structural design profession does consider earthquake induced forces regularly in the design practice with a return period of 2,500 years (NBCC-2005, IBC-2000). The devastation that may be caused by a tsunami of a large magnitude can be catastrophic as demonstrated by the 2004 Indian Ocean event which induced significant structural damage on infrastructure, killing over 300,000 people and leaving an estimated 1.5 million homeless (Ghobarah et al., 2006).

M. Saatçioğlu (✉)
Department of Civil Engineering, University of Ottawa, Ottawa, Canada
e-mail: murat.saatcioglu@uottawa.ca

A.T. Tankut (ed.), *Earthquakes and Tsunamis,* Geotechnical, Geological, and Earthquake Engineering 11, DOI 10.1007/978-90-481-2399-5_10,

The design of structures in flood-prone areas has previously been investigated. Design guidelines have been developed and codified, containing prescriptive provisions for flood induced loads (UBC 1997, ASCE 7-05 2005, IBC 2006). These codes may be used for the design of structures subjected to coastal flooding due to storm surges and flooding of river banks above bank-full conditions. In contrast, the design of structures against tsunami forces has received limited coverage in codes and standards. Recent research indicates that these forces can be significantly higher than those caused by wind storms, and can be comparable or in excess of those caused by earthquakes (Saatcioglu et al. 2006a, Nouri et al. 2007). At present, four design codes and guidelines specifically address tsunami-induced loads. These include; (i) The US Federal Emergency Management Agency recommendations (FEMA 55 2003) for tsunami-induced flood and wave loads, (ii) The City and County of Honolulu Building Code (2000) developed by the Department of Planning and Permitting of Honolulu, Hawaii, US, for districts located in flood and tsunami-risk areas, (iii) Structural Design Method of Buildings for Tsunami Resistance (SMBTR) proposed by the Building Center of Japan (Okada et al. 2005) outlining structural design for tsunami refuge buildings, and (iv) Guidelines for Structures that Serve as Tsunami Vertical Evacuation Sites, prepared by Yeh et al. (2005) for the US State Department of Natural Resources to estimate tsunami-induced forces on structures.

This paper demonstrates structural and non-structural damage observed during a reconnaissance visit to the tsunami stricken regions of Thailand and Indonesia after the 2004 Indian Ocean Tsunami and presents existing approaches for the computation of structural design loads generated by tsunamis.

2 Performance of Structures During the 2004 Indian Ocean Tsunami

The reconnaissance visit to Thailand covered; (i) the island of Phuket, (ii) Phi-Phi island about 48 km south east of Phuket, and (iii) the coastal town of Khao Lak about 100 km north of Phuket. The visit to Indonesia focused on the coastal city of Banda Aceh on the island of Sumatra.

The impact of tsunami was a function of the topography of coastal areas. In flat areas of coastal Sumatra, including the city of Banda Aceh, the tsunami waves reached 4–5 km inland, affecting a large population. In these areas the maximum wave height was about 4 to 6 m. The water mark on buildings along the southern coast of Phuket was measured to vary between 4 and 6 m from the sea level. Further north on Phuket Island and on Khao Lak Beach the water height was in excess of 10 m, causing significant structural and non-structural damage. The water run-up in hilly terrains resulted in significantly higher water levels. A coastal engineering team from Japan measured the maximum tsunami run-up height in Rhiting, south west of Banda Aceh, to be 49 m (Shibayama 2005).

The filed investigation focused on urban areas that had engineered and non-engineered structures. Engineered structures were mostly in the form of reinforced concrete frame buildings with masonry infill walls. A large number of them in

Thailand were hotel buildings since areas most affected by the tsunami were popular tourist resorts. Both concrete block and clay brick masonry were used as non-structural elements. Non-engineered structures were in the form of low-rise reinforced concrete frames with masonry infill walls, confined masonry, or timber frame buildings. These types of construction were used predominantly for shops, hotels and residential accommodation.

2.1 Engineered Reinforced Concrete Frame Buildings

There were many low to mid-rise reinforced concrete frame buildings which appeared to have been engineered in the visited areas. These frame buildings survived the tsunami pressure without structural damage, though they suffered damage to non-structural elements, especially the first story masonry infill walls. Figure 1 shows examples of reinforced concrete hotel buildings in Thailand that survived tsunami loads without any sign of structural damage, although nearby non-engineered buildings were either partially or fully collapsed.

Although engineered frame buildings performed well, there were some exceptions to this observation in Nai Thon Beach, where water run-up effects due to the

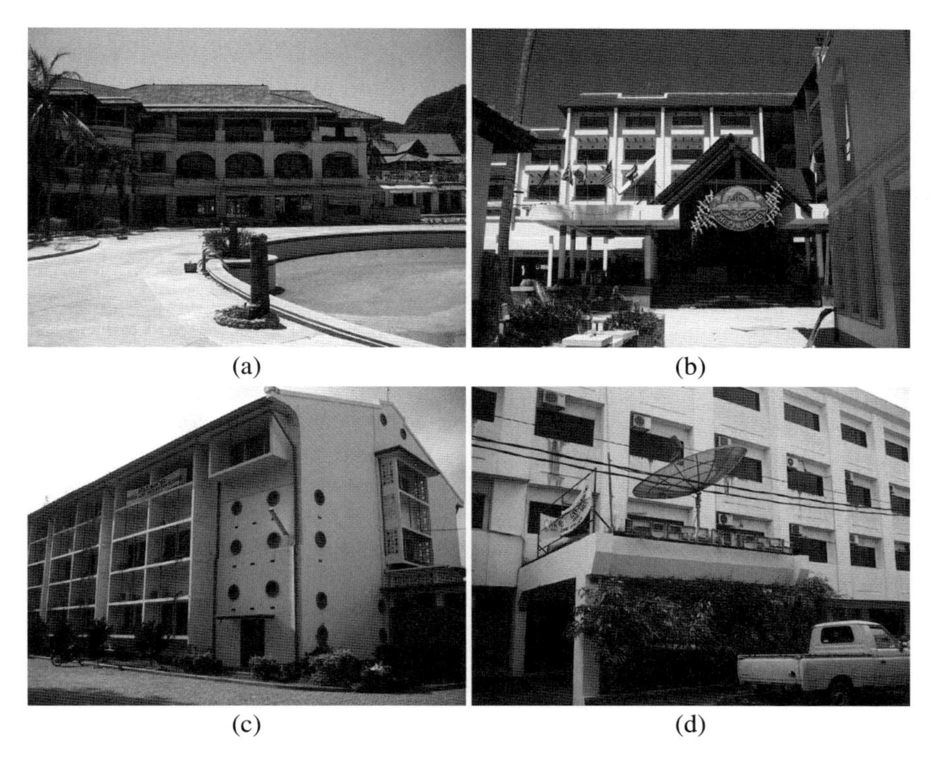

Fig. 1 Engineered reinforced concrete frame buildings, (**a**) and (**b**) on Phi Phi Island, Thailand; (**c**) and (**d**) in Banda Aceh, Indonesia

topography of the area between the ocean and the hilly terrain behind, resulted in increased water height. A large number of structural failures were observed in this region. Column failures due to insufficient flexural and shear capacities, as well as inadequate detailing of reinforcement to protect structures beyond the elastic range of deformations resulted in partial or full collapses of complete frames. Examples

(a) (b) (c) (d) (e) (f)

Fig. 2 Damage to reinforced concrete frame buildings in Nai Thon Beach, Thailand (See also Color Plate 15 on page 214)

of structural damage in Nai Thon Beach are illustrated in Fig. 2, which depicts poor column behavior.

Figure 2a, b show rear views of Khaolak Shopping Centre that suffered serious structural damage. The building was three stories high at the back, and a single story high facing the street at the front because of the grade. The tsunami waves hit the building from the ocean side at the back, covering the entire building with water up to the roof level as evidenced by the missing roofing tiles. The structural system consisted of reinforced concrete frames in the first two stories with slender circular columns, supporting a timber framing system at the third story level. The slender columns became unstable at the second story level, probably due to high secondary moments and failed. In some frames the continuity between the timber and concrete framing elements was poor and led to failures.

A reinforced concrete hotel complex, shown in Fig. 3 was under construction in Nai Thon Beach when it was hit by the tsunami. The frame elements, first storey stone masonry walls and second floor reinforced concrete slab were mostly completed at the time. Figure 3a shows the failure of the first storey masonry walls and the second floor shoring for the corner slab, exposing slab reinforcement in Fig. 3b. Figure 3c, d illustrate the failure of a corner column, which resulted in the failure of the two way slab system immediately above. Figure 3e shows the development of sufficiently wide flexural and inclined shear cracks in a spandrel beam, suggesting the yielding of reinforcement. It is clear that the tsunami pressure in the region was high enough to induce severe structural damage, all being triggered by the failure of weak columns, supporting deep and strong beams and attached two-way slab systems.

A common slab system that is used in Thailand consists of prefabricated reinforced concrete strips, supported by cast-in-place beams. These strips typically have 50 mm thickness, 300 mm width and 2.0 m length, reinforced with 4–6 mm diameter smooth wires, equally spaced in the centre of the section. Because of lack of proper connection to the supporting beams, these strips lifted up due to water pressure, causing slab failures. An example was a shopping centre in Patong Beach on Phuket Island where the below grade garage was filled up with water, lifting and destroying the first story slab panels as shown in Fig. 4a, b.

Other examples of damaged precast slab systems were observed in a number of different buildings in Thailand. Figure 4d shows a building in Nai Thon Beach that suffered dislocation of the precast slab system. A similar type of slab failure was also observed in the concrete dock of the Khao Lak Harbor in Thailand. The slab strips used in the harbor dock was thicker. The failure was in the form of lifting of individual strips due to water pressure, or lifting of the entire slab panel, as shown in Fig. 5 when the strips had cast-in-place topping.

2.2 Non-Engineered Reinforced Concrete Frame Buildings

The majority of one to two story low-rise buildings were constructed using site-cast concrete without much evidence of engineering design. The columns were

Fig. 3 Damage to reinforced concrete frame building under construction in Nai Thon Beach, Thailand

of very small cross-section (about 200 mm square), containing 4–8 mm diameter smooth or deformed corner bars, resulting in approximately 0.5% reinforcement ratio. The column ties were widely spaced, with little shear resistance. Column failures were blamed for the collapse of entire structural systems in non-engineered low-rise reinforced concrete buildings when the tsunami water height was 4–6 m

Fig. 4 (**a**) (**b**) Damage to precast slab strips in Patong Beach shopping centre in Thailand; (**c**) Close-up view of a precast slab strip; (**d**) lifting of precast concrete slab due to water pressure in Nai Thon Beach, Thailand

and the columns received significant lateral pressure due to the attached panels. Figure 6 shows partial and full structural collapses in Thailand and Indonesia. However, many similar columns survived the tsunami without much damage,

Fig. 5 Damage to precast slab strips of the concrete dock in Khao Lak, Thailand

especially those that were away from the shore and those that had additional lateral bracings provided by in-plane infill walls. Figure 7 illustrates non-engineered frame buildings that remained intact during the tsunami, some with damaged non-structural walls.

2.3 Unreinforced Masonry Infill Walls and Confined Masonry

The majority of buildings in Thailand and Indonesia had reinforced concrete or timber frames infilled with unreinforced masonry walls, or confined masonry with lightly reinforced concrete horizontal and vertical tie elements (beams and columns). The masonry units used were consistently of the same type, with 50 mm thickness. Both hallow clay bricks and concrete masonry blocks of the same thickness were used as illustrated in Fig. 8. This figure also illustrates a typical confined masonry building during construction.

The masonry walls suffered extensive out of plane failures when subjected to tsunami pressures perpendicular to the wall plane. The water pressure resulted in large holes in lower story walls, sometimes removing the masonry almost entirely. The remains of walls around the frames did not show any sign of diagonal cracking, unlike those observed under seismic excitations. Figure 9 shows typical punching failures observed in masonry infill walls. When these failures occurred in load bearing walls of confined masonry, partial or total collapses were observed, as shown in Fig. 10.

2.4 Timber Construction

Low-rise timber construction, shown in Fig. 11, is typically used for low cost and affordable housing in Banda Aceh. The roofs either have light corrugated metal coverage or clay tiles. These houses appeared to have survived the earthquake that caused the tsunami and hit the area, with minor damage, but many collapsed completely under tsunami wave pressures. The building components disintegrated into smaller pieces, contributing to debris that impacted on other structures. Figure 12 illustrates the devastation due to massive failures of timber buildings in residential areas. Some of the very same timber buildings did survive the tsunami, especially if they were away from the coast, protected by surrounding buildings and laterally braced by masonry walls.

2.5 Lack of Anchorage

The tsunami pressure and wave height were sufficient to displace structures and structural components from their foundations and float them hundreds of meters away. Figure 13 shows a single story reinforced concrete framed house with masonry

(a) (b) (c) (d)

Fig. 6 Damage to non-engineered reinforced concrete; (**a**) Phuket, Thailand (**b**) Phi-Phi Island, Thailand, (**c**) and (**d**) Banda Aceh, Indonesia

(a) (b)

Fig. 7 Non-engineered concrete frames that survived the tsunami; (**a**) Khao Lak Beach, Thailand and (**b**) Banda Aceh, Indonesia

walls that was displaced and floated away from its foundation, resting on the nearby street. The lack of anchorage was attributed to poor and shallow foundation on soft soil.

Fuel supplies in Banda Aceh were disrupted due to damage to oil storage tanks. The fuel storage tank depot suffered extensive damage from the tsunami. The retreating water displaced three large tanks a distance of approximately half a kilometer as shown in Fig. 14. The tanks were torn from their connecting pipes, lost their fuel content and impacted and damaged many houses as seen in Fig. 14. Lack of anchor bolts, which is the common practice for these tanks, was blamed for the failure.

(a) (b)

Fig. 8 (**a**) Typical 50 mm thick concrete block and clay brick masonry units; (**b**) construction of confined masonry in Khao Lak harbor town in Thailand

(a) (b)

Fig. 9 Typical punching failure of masonry walls; (**a**) Phi Phi Island, Thiland; (**b**) Banda Aceh, Indonesia

(a) (b)

Fig. 10 Failure of confined masonry in Banda Aced, Indonesia

2.6 Tsunami Loads

Two types of tsunami related loads were applied on structures; (i) tsunami water pressure (hydrostatic, buoyant, hydrodynamic and surge), and (ii) the impact of floating debris. Tsunami waves imposed water pressures due to the impulse of breaking waves along the shore, and dynamic pressures that varied with water celerity and height. The impulse component diminished as water moved inland. The hydrodynamic pressure also decreased due to reduced water velocity caused by surface friction.

It became evident during the field investigation that some of the damage was caused by floating debris, especially in Banda Aceh, Indonesia. The flat topography of Banda Aceh and sufficient water height and pressure allowed not only the remains of collapsed building components to float but also large and heavy objects; such as cars, trucks and fishing vessels to float and impact on the physical infrastructure. It was clear that many small size building columns were damaged significantly by impact forces caused by floating debris, resulting in the collapse of slabs supported

(a) (b)

Fig. 11 Timber framed structures in Banda Aced, Indonesia

(a) (b)

Fig. 12 Destruction of entire residential districts of timber framed structures in Banda Aceh, Indonesia

by such critical elements. Figures 15 and 16 illustrate the impact loading due to debris.

A unique aspect of Banda Aceh, from impact loading perspective, was the presence of a large number of fishing boats which were spread around various harbors, as well as rivers that pass through the city. These boats floated during the tsunami, impacting on buildings as illustrated in Fig. 16. A large floating power generator, the same size and appearance of a large size oil tanker docked at the harbor was freed up by tsunami forces and floated 3.5 km inland, destroying residential buildings on

Fig. 13 A single-story house, displaced by water pressure due to lack of proper anchorage, Banda Aceh, Indonesia (See also Color Plate 16 on page 215)

(a) (b)

Fig. 14 A single-story house, displaced by water pressure due to lack of proper anchorage, Banda Aceh, Indonesia

its way. This is shown in Fig. 17 in its current location, which happens to be on top of several residential buildings that were completely crushed.

3 Tsunami Forces on Structures

A broken tsunami wave generates forces which affect structures located in its path. A comprehensive review of tsunami forces is presented by Nistor et al. (2008), and is reproduced in this section. Three parameters are essential for defining the magnitude and application of these forces: (1) inundation depth, (2) flow velocity, and (3) flow direction. The parameters mainly depend on: (a) tsunami wave height and wave period; (b) coastal topography; and (c) roughness of the coastal inland. The extent of tsunami-induced inundation depth at a specific location can be estimated using various tsunami scenarios (magnitude and direction) and coastal inundation modeling. However, the flow velocity and direction is generally more difficult to estimate. Flow velocities vary in magnitude from zero to significantly high values, while flow direction varies due to onshore topography, as well as soil cover and obstacles. Forces associated with tsunami bores consist of: (1) hydrostatic force, (2) hydrodynamic (drag) force, (3) buoyant force, (4) surge force and (5) impact of debris. A brief description of these forces is presented below.

3.1 Hydrostatic Force

The hydrostatic force is generated by still or slow-moving water acting perpendicular onto planar surfaces. The hydrostatic force per unit width, F_{HS}, can be calculated using the equation given below.

(a) (b) (c) (d)

Fig. 15 Impact loading on columns due to floating debris, Banda Aceh, Indonesia (See also Color Plate 17 on page 215)

$$F_{HS} = \frac{1}{2}\rho g \left(d_s + \frac{u_p^2}{2g}\right)^2 \qquad (1)$$

Where ρ is the seawater density, g is the gravitational acceleration, d_s is the inundation depth and u_p is the normal component of flow velocity. Equation (1) is proposed by the City and County of Honolulu Building Code (CCH) and accounts for the velocity head. Alternatively, FEMA 55 does not include the velocity head in its formulation since it is assumed to be a negligible component of the hydrostatic force.

The point of application of the resultant hydrostatic force is located at one third from the base of the triangular hydrostatic pressure distribution. In the case of a broken tsunami wave, the hydrostatic force is significantly smaller than the drag and surge forces. However, the hydrostatic force can be important when the effect of tsunami is similar to a rapidly-rising tide (Dames and Moore 1980).

Fig. 16 Impact of fishing boats on buildings, Banda Aceh, Indonesia

Fig. 17 Power generating vessel that floated 3.5 km inland in Banda Aceh, Indonesia

3.2 Buoyant Force

The buoyant force is the vertical force acting through the center of mass of a submerged body. Its magnitude is equal to the weight of the volume of water displaced by the submerged body. The effect of buoyant forces generated by tsunami flooding was clearly evident during the 2004 Indian Ocean Tsunami field observations, as illustrated in Figs. 4 and 5, where reinforced concrete slabs were lifted up and displaced due to buoyant forces. Buoyant forces can generate significant damage to structural elements, and are calculated as follows:

$$F_B = \rho g V \tag{2}$$

Where, V is the volume of water displaced by submerged structure.

3.3 Hydrodynamic (Drag) Force

As the tsunami bore moves inland with moderate to high velocity, structures are subjected to hydrodynamic forces caused by drag. Currently, there are differences in estimating the magnitude of the hydrodynamic force. The general expression used by existing codes is given below.

$$F_D = \frac{\rho C_D A u^2}{2} \tag{3}$$

Where, F_D is the drag force acting in the direction of flow and C_D is the drag coefficient that depends on the shape of the surface on which drag forces are applied. Drag coefficient values of 1.0 and 1.2 are recommended for circular piles by CCH and FEMA 55, respectively. For the case of rectangular piles, the drag coefficient recommended by the same two codes is 2.0. Parameter A is the projected area of the body normal to the direction of flow. The flow is assumed to be uniform, and therefore, the resultant force will act at the centroid of the projected area A. The term u in Eq. (3) is the tsunami bore velocity. The hydrodynamic force is directly proportional to the square of the tsunami bore velocity as indicated in Eq. (3). Therefore, the estimation of the bore velocity remains to be one of the critical elements on which there is significant disagreement in literature.

Tsunami-bore velocity and direction can vary significantly during a major tsunami inundation. Current estimates of the velocity are crude. A conservatively high flow velocity impacting the structure at a normal angle may be assumed to estimate a conservative value for the drag force. Also, the effects of run-up, backwash and direction of velocity are not addressed in current design codes. Although there is certain consensus in the general form of the equation for the bore velocity, several researchers proposed different empirical coefficients. The general form of the bore velocity is shown below.

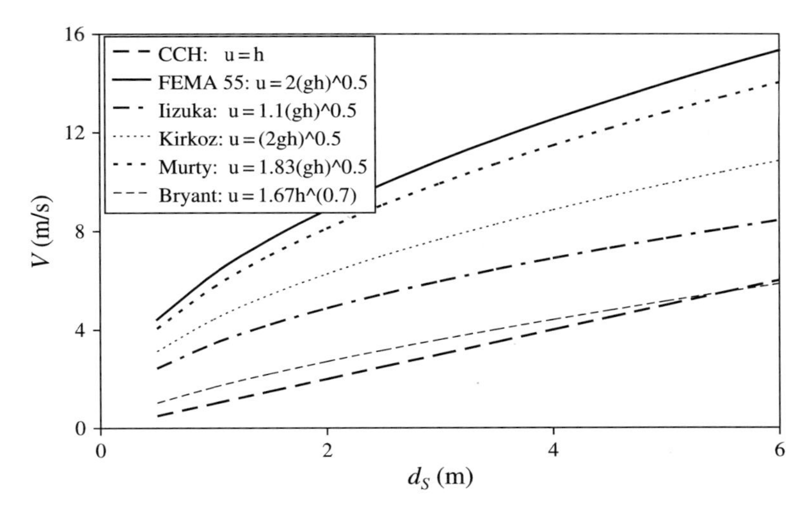

Fig. 18 Comparison of various tsunami-bore velocities as a function of inundation depth (Nouri et al., 2007)

$$u = C\sqrt{gd_s} \tag{4}$$

Where, C is a constant and d_s is the inundation depth. Various formulations were proposed by FEMA 55 (2003) based on Dames and Moore (1980), Iizuka and Matsutomi (2000), CCH (2000), Kirkoz (1983), Murty (1997), Bryant (2001), and Camfield (1980) for estimating the velocity of a tsunami bore in terms of inundation depth. The velocity inundation depth relationships proposed are plotted in Fig. 18. It can be seen that bore velocities calculated using CCH (2000) and FEMA 55 (2003) represent lower and upper bound values, respectively.

3.4 Surge Force

The surge force is generated by the impingement of the advancing water front of a tsunami bore on a structure. Due to lack of detailed experimental research specifically applicable to tsunami bores running up the shoreline, the calculation of the surge force exerted on a structure is subject to substantial uncertainty. Based on research conducted by Dames and Moore (1980), CCH (2000) recommends the following expression for the computation of surge force F_s on walls with heights equal to or greater than three times the surge height (*3h*).

$$F_S = 4.5\rho g h^2 \tag{5}$$

Where, F_s is the surge force per unit width of wall, and h is the surge height. The point of application of the resultant surge force is located at a distance h above the base of the wall. Structural walls with heights less than *3h* require surge forces to be calculated using an appropriate combination of hydrostatic and drag forces for each specific situation. SMBTR recommends using the equation for tsunami wave pressure derived by Asakura et al. (2000), as given in Eq. (6). The equivalent static pressure resulting from the tsunami impact is associated with a triangular distribution where water depth equals three times the tsunami inundation depth.

$$qx = \rho g(3h - z) \tag{6}$$

Where, qx is the tsunami wave pressure for structural design, z is the height of the relevant portion from ground level $(0 \leq z \leq 3h)$, ρ is the mass per unit volume of water and g is the gravitational acceleration.

Integration of the wave pressure formula for walls with heights equal to or greater than $3h$ results in the same equation as the surge force formula recommended by CCH (Eq. (5)). The magnitude of the surge force calculated using Eqs. (5) and (6) generate a value equal to nine times the magnitude of the hydrostatic force for the same flow depth. However, experimental research conducted by Ramsden (1996), Arnason (2005) did not indicate such differences in magnitude. Yeh et al. (2005) commented on the validity of Eq. (5) and indicated that this equation gives "excessively overestimated values". On the other hand, Nakano and Paku (2005) conducted extensive field surveys in order to examine the validity of the proposed tsunami wave pressure formula given in Eq. (6). The factor 3.0 in Eq. (6) was taken as a variable, α, and was calculated such that it could represent the boundary between damage and no damage in the surveyed data. A value of α equal to 3.0 and 2.0 was found for walls and columns, respectively. The former is in agreement with the proposed formulae by both CCH and SMBTR (Eqs. (5) and (6)). The tsunami wave force may be composed of drag, inertia, impulse and hydraulic gradient components. However, SMBTR does not specify different components for the tsunami-induced force, and the proposed formula presumably accounts for other components.

3.5 Debris Impact Force

A high-speed tsunami bore traveling inland carries debris such as floating automobiles, floating pieces of buildings, drift wood, boats and ships. The impact of floating debris can induce significant forces on a building, leading to structural damage or collapse (Saatcioglu et al. 2006b, c). This is illustrated in Figs. 15, 16, and 17. Both FEMA 55 (2003) and CCH (2000) codes account for debris impact forces, using the same approach, and recommend using Eq. (7) for estimation of debris impact force.

Table 1 Impact duration of floating debris (FEMA 55)

	Impact duration (seconds)	
Type of construction	Wall	Pile
Wood	0.7–1.1	0.5–1.0
Steel	N.A.	0.2–0.4
Reinforced concrete	0.2–0.4	0.3–0.6
Concrete masonry	0.3–0.6	0.3–0.6

$$F_i = m_b \frac{du_b}{dt} = m \frac{u_i}{\Delta t} \tag{7}$$

Where F_i is the impact force, m_b is the mass of the body impacting the structure, u_b is the velocity of the impacting body (assumed equal to the flow velocity), u_i is approach velocity of the impacting body (assumed equal to the flow velocity) and Δt is the impact duration taken equal to the time between the initial contact of the floating body with the building and the instant of maximum impact force. The only difference between CCH and FEMA 55 resides in the recommended values for the impact duration which has a noticeable effect on the magnitude of the force. For example, CCH recommends the use of impact duration of 0.1 s for concrete structures, while FEMA 55 provides different values for walls and piles for various construction types as shown in Table 1.

According to FEMA 55 (2003), the impact force (a single concentrated load) acts horizontally at the flow surface or at any point below it. Its magnitude is equal to the force generated by 455 kg (1,000-pound) of debris traveling with the bore and acting on a 0.092 m^2 (1 ft^2) surface of the structural element. The impact force is to be applied to the structural element at its most critical location, as determined by the structural designer. It is assumed that the velocity of the floating body goes from u_b to zero over some small finite time interval Δt. Finding the most critical location of impact is a trial and error procedure that depends, to a large extent, on the experience and intuition of the engineer.

3.6 Breaking Wave Forces

Tsunami waves tend to break offshore and approach shoreline as a broken hydraulic bore or a soliton, depending on wave characteristics and coastal bathymetry. Consequently, classic breaking wave force formulas are not directly applicable to the case of tsunami bores. Hence, this paper does not discuss the estimation of breaking wave forces.

3.7 Load Combinations for Calculating Tsunami-Induced Forces

Appropriate combinations of tsunami-induced force components (hydrostatic, hydrodynamic, surge, buoyant and debris impact force) should be used in calculating the total tsunami force, based on the location and type of structural elements under consideration. This is due to the fact that a certain element may not be subjected to all of these force components simultaneously. Loading combinations can significantly influence the total tsunami force and the subsequent structural design. Unlike the case of tsunami waves, loading combinations for flood-induced surges are well-established and have been included in design codes. The literature review revealed that proposed tsunami load combinations must be significantly improved and incorporated in new design codes. Tsunami-induced loads are different from flood-induced loads. Therefore, load combinations based on flood surges are not directly applicable to tsunamis. Loading combinations proposed in the literature are as follows:

FEMA 55 (2003) does not provide loading combinations specifically for the calculation of tsunami force. However, flood load combinations can be used as guidance. Flood load combinations for piles or open foundations, as well as solid walls (foundation) in flood hazard zones and coastal high hazard zones are presented as follows. F_{brkp}, F_{brkw}, F_i and F_{dyn} refer to breaking force on piles, breaking force of walls, impact force and hydrodynamic force, respectively.

Pile or Open Foundation:
F_{brkp} (on all piles) + F_i (on one corner or critical pile only), or
F_{brkp} (on front row of piles only) + F_{dyn} (on all piles but front row) + F_i (on one corner or critical pile only).

Solid (Wall) Foundation:
F_{brkw} (on walls facing shoreline, including hydrostatic component) + F_{dyn} (assumes one corner is destroyed by debris).

Yeh et al. (2005) modified flood load combinations provided by FEMA 55 and adapted them for tsunami forces as follows:

Pile or Open Foundation:
F_{brkp} (on column) + F_i (on column), or
F_d (on column) + F_i (on column).
Where F_d is the drag force.

Solid (Wall) Foundation (Perpendicular to Flow Direction):
F_{brkw} (on walls facing shoreline) + F_i (on one corner), or
F_s (on walls facing shoreline) + F_i (on one corner), or
F_d (on walls facing shoreline) + F_i (on one corner).
where F_s is the surge force on walls.

Dias et al. (2005) proposed two load combinations called "point of impact" and "post-submergence/submerged". These load combinations are based on two conditions: (i) the instant that tsunami-bore impacts the structure, and (ii) when the whole structure is inundated.

Point of Impact:
F_d (on walls facing shoreline) + F_s (on walls facing shoreline).
where F_s is defined as the hydrostatic force by Dias et al. (2005).

Post-Submergence/Submerged:
F_d (on walls facing shoreline) + F_b (on submerged section of the structure).

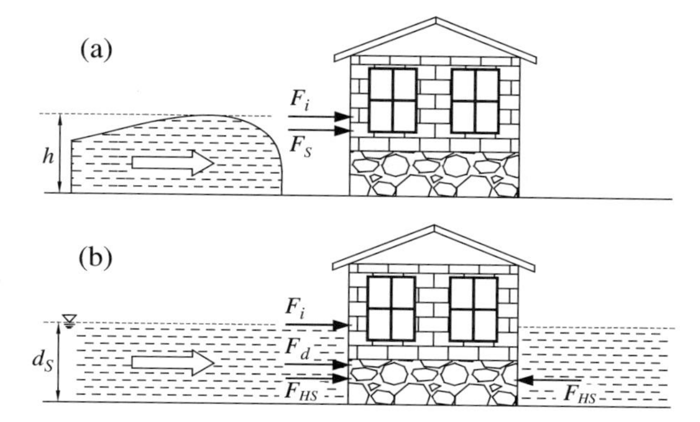

Fig. 19 Proposed loading conditions: (**a**) point of impact; and (**b**) post-impact, (Nouri et al. 2007)

The net hydrostatic force is zero and F_b is the buoyant force.

Nouri et al. (2007) proposed two new load combinations based on the two conditions considered by Dias et al. (2005), as shown in Fig. 19. The proposed load combinations by Nouri et al. (2007) are adapted to follow a consistent format as the above combinations:

Columns:
F_s (on front row of piles only) + F_i (on one corner or critical column in the front row only), or
F_d (on all piles) + F_i (on one corner or critical column only).
where F_s is the surge force on walls.

Solid (wall) Foundation:
F_s (on walls facing shoreline) + F_i (on walls facing shoreline), or
F_d (on walls facing shoreline) + F_i (on one critical wall facing shoreline) + F_b (on submerged section of the structure).

4 Case Study – 10-Sorey Reinforced Concrete Office Building

A 10-storey reinforced concrete shear wall building in Vancouver, Canada is considered for comparison of earthquake and tsunami-induced lateral forces. Figure 20 illustrates the elevation view of the short side of the building, exposed to tsunami forces, and a typical plan view, including cross-sectional details of structural elements. The structure is an office building oriented such that the short side is parallel to the shoreline. It is to be designed for earthquake forces, as per the National Building Code of Canada (NBCC-2005), as well as tsunami forces following the CCH (2000) Building Code. The tsunami inundation height is assumed to be 5.0 m, which is the level typically observed in Banda Aced, Indonesia and Phuket, Thailand during the 2004 Indian Ocean Tsunami. It is assumed that the maximum likely mass

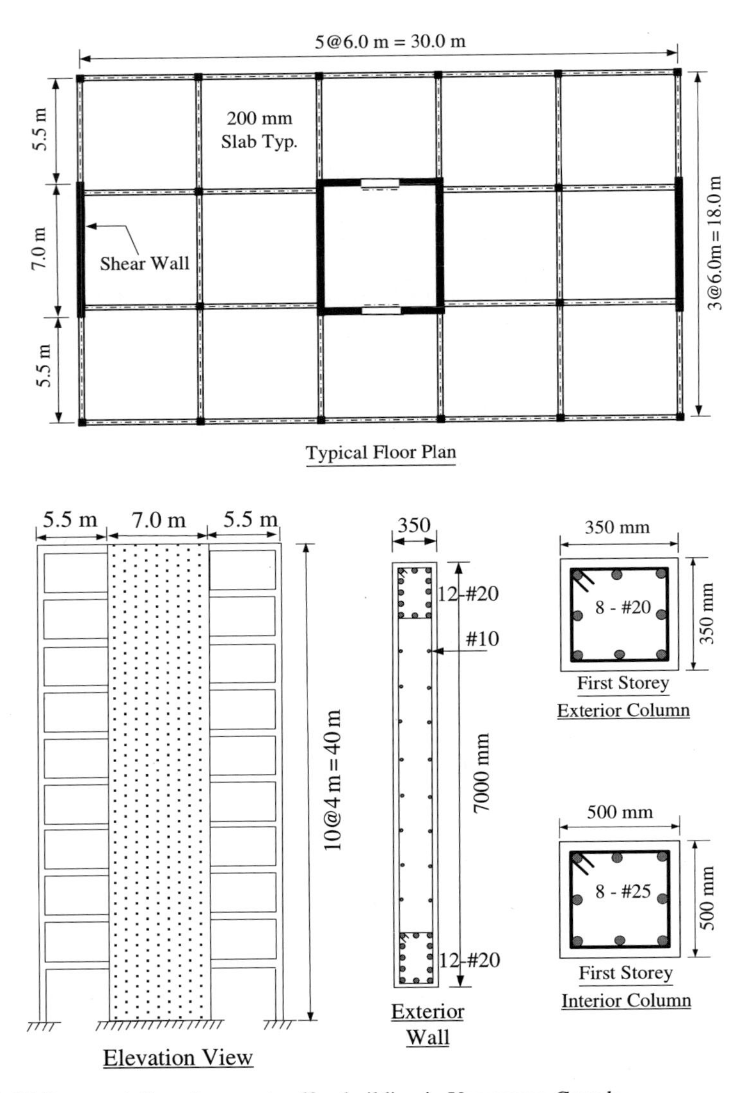

Fig. 20 10-Storey reinforced concrete office building in Vancouver, Canada

causing debris impact force is about 1,000 kg. The building is located on firm soil and does not have any exterior cladding that is attached to the structural system to contribute towards the tributary area of columns other than glass windows which are assumed to fail immediately under tsunami generated surge forces. Use sea (salt) water density of $\rho = 1030\,\text{kg/m}^3$, $g = 9.81\,\text{m/s}^2$, concrete compressive strength of 40 MPa and reinforcement yield strength are 400 MPa. Slab dead load (DL) and live load (LL) are; $5.0\,\text{kN/m}^2$ and $2.4\,\text{kN/m}^2$ for typical floors and $3.5\,\text{kN/m}^2$ and $2.2\,\text{kN/m}^2$ for the roof, respectively.

Surge Force: $F_s = 4.5\rho g h^2 = 4.5(1030)(9.81)(5.0)^2 = 1.14 \times 10^6\ N/m$
$F_s \times (surface\ width) = (1.14 \times 10^6)(2 \times 0.35 + 7.0) = 8{,}778 \times 10^3\ N$

Drag Force: $F_D = \dfrac{\rho C_D A u^2}{2}$; where $u = d_s = 5\,m/s$ and $C_D = 2.0$

$$A = (5.0)(12 \times 0.35 + 4 \times 0.5 + 4 \times 7.0) = 171\,m^2$$

$$F_D = \frac{(1030)(2.0)(171)(5.0)^2}{2} = 4{,}403 \times 10^3\ N$$

Debris Impact Force: $F_i = m\dfrac{u_i}{\Delta t} = 1000\dfrac{5.0}{0.1} = 50 \times 10^3\ N$

Load Combination: Use the combinations suggested by Nouri (2007);
Surge + Impact = 8, 778 + 50 = 8, 828 kN
Drag + Impact = 4, 403 + 50 = 4, 453 kN
Governing Tsunami Force = 8,828 kN

Seismic Base Shear Force as per NBCC-2005: $V = \dfrac{S(T_a)M_v I W}{R_d R_o}$

$T_a = 0.05(h_n)^{3/4} = 0.05(40)^{3/4} = 0.8 sec$ *for shearwallbuildings*
$S(0.8) = 0.454$ *for firm soil in Vancouver, Canada*
$M_v = 1.0$ *for* $T_a < 1.0\,s$
$R_d = 3.5$ *and* $R_o = 1.6$ *for ductile shearwall buildings*
$I = 1.0$ *for office buildings*
$W = 37{,}750\,kN$ *(structural weight* $= DL + 0.25LL$*)*
$V_{elastic} = (0.454)(1.0)(1.0)(37{,}750) = 17{,}138\,kN$

$$V_{inelastic} = \frac{V_{elastic}}{R_d R_o} = \frac{17{,}138}{(3.5)(1.6)} = 3{,}060\,kN$$

The comparison shows that the total tsunami design force of 8,828 kN, as total base shear is higher than the inelastic seismic design base shear of 3,060 kN. Therefore, the structure as designed for seismic forces appears to be deficient against tsunami forces. However, if the building is designed for elastic seismic response, then the building clearly has sufficient capacity against tsunami forces.

Unlike earthquake induced inertia forces that affect the entire lateral load resisting system, allowing redistribution of forces and stresses in a ductile system, tsunami pressures may be more critical on local elements. Therefore, the effect of tsunami loads on individual elements has also been investigated, as shown below.
Tsunami Forces on Columns: Consider a first-storey corner column with an unfactored axial gravity load of P = 478 kN and see Fig. 21.

Surge Force:
$F_{s1} = 11\rho g(4)(0.35) = 11(1030)(9.81)(4)(0.35) = 156 \times 10^3\ N$
$F_{s2} = 4\rho g(4)(0.35)/2 = 4(1030)(9.81)(4)(0.35)/2 = 28 \times 10^3\ N$
Drag Force: $F_D = \dfrac{\rho C_D A u^2}{2} = \dfrac{1030(2.0)(4.0 \times 0.35)(5)^2}{2} = 36 \times 10^3\ N$

Debris Impact Force: $50 \times 10^3\ N$ (assumed to act at the location of F_{s1})

Use the combination of the governing Surge and Debris Impact Force. Assuming full fixity at the footing and the first-storey level, the maximum column moment of $M_{base} = 120\,kN.m$ occurs at the base. The nominal column capacity for a 350 mm square column with 8-#20 (20 mm diameter) bars is computed to be, $M_n = 198\,kN.m > M_{base} = 120\,kN.m$ indicating that corner columns are safe against tsunami forces.

Tsunami Forces on Walls: Consider a first-storey exterior shear wall, fully exposed to tsunami pressures with an unfactored axial gravity load of $P = 1,143\,kN$.

Surge Force:
$F_{s1} = 11\rho g(4)(7.0) = 11(1030)(9.81)(4)(7.0) = 3112 \times 10^3\ N$
$F_{s2} = 4\rho g(4)(7.0)/2 = 4(1030)(9.81)(4)(7.0)/2 = 566 \times 10^3\ N$
Drag Force: $F_D = \dfrac{\rho C_D A u^2}{2} = \dfrac{1030(2.0)(4.0 \times 7.0)(5)^2}{2} = 721 \times 10^3\ N$
Debris Impact Force: $50 \times 10^3\ N$ (assumed to act at the location of F_{s1})

Use the combination of the governing Surge and Debris Impact Force. Assuming bending about weak axis and full fixity at the footing and the first-storey level, the maximum wall moment occurs at the base; $M_{base} = 1,892\,kN.m$. The nominal wall capacity about the weak axis is computed to be $M_n = 515\,kN.m < M_{base} =$

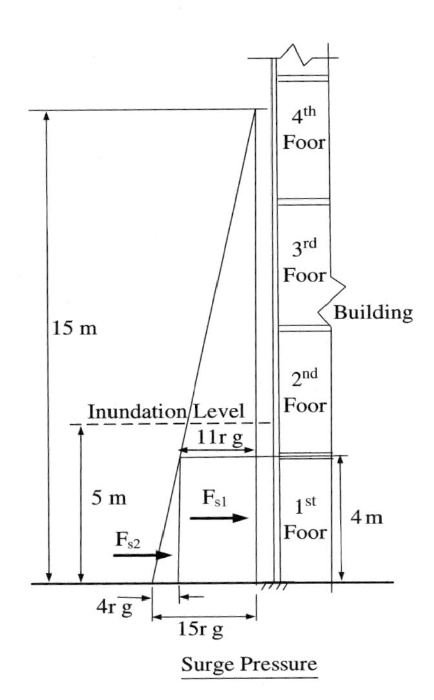

Fig. 21 Tsunami surge pressure on the 10-storey building

1, 892 kN.m indicating that exterior shear wall does not have sufficient flexural capacity against the computed tsunami design force.

5 Conclusions

Performance of buildings during the 2004 Indian Ocean Tsunami indicate that tsunami generated hydraulic pressures and debris impact forces can have significant effects on structural behaviour. Engineered structures may survive tsunami forces, especially if designed for such forces. However, non-engineered buildings, including reinforced concrete frames with infill masonry walls, confined masonry and timber framed buildings may be at high risk depending on the characteristics of tsunami wave pressures and debris impact forces. Designing buildings for earthquake loads may not ensure safety under large magnitude tsunami events. Though some design provisions are included in selected building codes, further research is needed to improve the accuracy of tsunami load predictions, as well as design and detailing requirements for structural and non-structural components.

References

Asakura, R., Iwase, K. and Iketani, T. (2000): *Proceedings of the Coastal Engineering of JSCE*, Vol. 47, pp. 911–915.

ASCE Standard (2006): Minimum design loads for buildings and other structures. SEI/ASCE 7-05, p. 424.

Bryant, E. A. (2001): *Tsunami: the Underrated Hazard*, Cambridge University Press, London, UK, p. 320.

Camfield, F. (1980): *Tsunami Engineering*, Coastal Engineering Research Center, US Army Corps of Engineers, Special Report (SR-6), p. 222.

Dames and Moore (1980): Design and Construction Standards for Residential Construction in Tsunami-Prone Areas in Hawaii, (Prepared for the Federal Emergency Management Agency).

Dias, P., Fernando, L., Wathurapatha, S. and De Silva, Y. (2005): *International Conference of Disaster Reduction on Coasts*, Melbourne.

Federal Emergency Management Agency (2003): *Coastal Construction Manual* (3 vols.), 3rd Ed. (FEMA 55), Jessup, MD.

Ghobarah, A., Saatcioglu, M. and Nistor, I. (2006): *Engineering Structures*, Vol. 28, pp. 312–326.

HHC (2000): Department of Planning and Permitting of Honolulu Hawai, Chapter 16, City and County of Honolulu Building Code, *Article 11*.

IBC (2006): International Code Council (ICC), *International Building Code 2006*, Country Club Hills, IL, p. 675.

Iizuka, H. and Matsutomi, H. (2000): *Proc. Conf. Coastal Engrg.*, JSCE, 47. (in Japanese).

Kirkoz, M.S. (1983): *10th IUGG International Tsunami Symposium*, Sendai-shi/Miyagi-ken, Japan, Terra Scientific Publishing, Tokyo, Japan.

Murty, T.S. (1997): Bulletin of the Fisheries Research Board of Canada- no. 198, Department of Fisheries and the Environment, Fisheries and Marine Service, Scientific Information and Publishing Branch, Ottawa, Canada.

Nakano, Y. and Paku, C. (2005): Summaries of technical papers of Annual Meeting Architectural Institute of Japan (Kinki).

NBCC (2005): National Building Code of Canada, National Research Council of Canada, Ottawa, Canada.

Nistor, I., Palermo, D., Nouri, Y., Murty, T. and Saatcioglu, M. (2008): Tsunami Induced Forces on Structures, Chapter 11 of *Handbook of Coastal and Ocean Engineering*.

Nouri, Y., Nistor, I., Palermo, D. and Saatcioglu, M. (2007): *9th Canadian Conference of Earthquake Engineering*, Ottawa, Canada.

Ramsden, J.D. (1996): *Waterways, Port Coasts and Ocean Eng.*, 122(3), pp. 134–141.

Saatçioğlu, M., Ghobarah, A. and Nistor, I. (2006a): "Effects of the December 26, 2004 Sumatra Earthquake and Tsunami on Physical Infrastructure," *ISET Journal of Earthquake Engineering*.

Saatçioğlu, M., Ghobarah, A. and Nistor, I. (2006b): *Earthquake Spectra*, Earthquake Engineering Research Institute, 22, pp. 295–320.

Saatçioğlu, M., Ghobarah, A. and Nistor, I.(2006c): *Earthquake Spectra*, Earthquake Engineering Research Institute, 22, pp. 355–375.

Shibayama, T. (2005): Yokohama National University, http://www.cvg.ynu.ac.jp/G2/indonesia_survey_ynu_e.html, Accessed March 2005.

UBC (1997): *International Conference of Building Officials*. 1997 Uniform Building Code. California.

Yeh, H., Robertson, I. and Preuss, J. (2005): *Report No 2005-4*, Washington Dept. of Natural Resources.

Characteristics of Disasters Induced by the Wenchuan 8.0 Earthquake and Its Lessons

Wang Lanmin, Wu Zhijian, and Sun Junjie

Abstract On May 12, 2008, a great earthquake with a magnitude 8.0 occurred in the Wenchuan County, Sichuan Province, China. The main shock was widely felt by the people in 30 provinces of China, 6 of which were seriously affected, and 3 of them were seriously damaged. The strong shaking not only caused enormous buildings and houses collapsed or seriously damaged, but also triggered more than 12,000 landslides, collapses and mudflows, which dammed more than 30 quake lakes. According to the authority statistics, 87,403 people were killed and more than 374,000 people injured. The direct economic loss reaches 845 billion Chinese Yuan, which is about 124 billion USD. In this paper, the characteristics of damage of buildings, houses, infrastructures and lifeline were described, which including reinforced concrete buildings, brick-concrete buildings, the bottom reinforced concrete buildings, brick-wood houses and adobe-wood houses, highways and railways, bridges, dams, electricity supply, water supply, gas supply and communication system. The secondary disasters induced by the earthquake, such as landslides, rolling rocks, seismic settlement and liquefaction were investigated and introduced. The characteristics and distribution of both ground motion and faults ruptures caused by the earthquake were presented, and the housing reconstruction conditions in quake-hit area were brief introduced as well. At last, some lessons learnt from the great earthquake were summarized.

Keywords Wenchuan · Earthquake · Disasters · Characteristics · Lessons

1 Introduction

On May 12, 2008 at 14:28, a great earthquake with a magnitude 8.0 occurred in the Wenchuan County, Sichuan Province in China. The Epicenter locates at the place of latitude N31.021 and longitude E103.367, and the focal depth is 14 Km.

W. Lanmin (✉)
Lanzhou Institute of Seismology, China Earthquake Administration, 450 Donggangxilu Ave., Lanzhou 730000, China
e-mail: Wanglm@gssb.gov.cn

A.T. Tankut (ed.), *Earthquakes and Tsunamis,* Geotechnical, Geological, and Earthquake Engineering 11, DOI 10.1007/978-90-481-2399-5_11,

Table 1 The distribution of casualties and economic loss in the Sichuan and the other regions around it in China

Region	Dead	Injured	Missing	Loss (10^9US$)
Total	69,209	374,498	18,194	130.87
Sichuan	68,696	360,236	18,194	121.31
Gansu	368	10,171		6.70
Shaaxi	122	3,379		2.86
Chongqing	18	637		
Yunnan	1	51		
Henan	2	7		
Hubei	1	17		
Hunan	1			

The earthquake caused tens of thousands of deaths and hundreds of billions RMB in losses, and it has become the worst earthquake event to occur since new China established in 1949, except the Ms7.8 Tangshan Earthquake in 1976. The distribution of casualties and economic loss in the Sichuan and the other regions around it is listed in Table 1. The event occurred along the Longmen fault, which has an up bound magnitude of 7.3 for potential seismic sources on the zonation map in Chin, and where the Diexi Earthquake with magnitude 7.5 occurred in 1933, and two other earthquakes both with magnitude of 7.2 occurred in the Songpan County

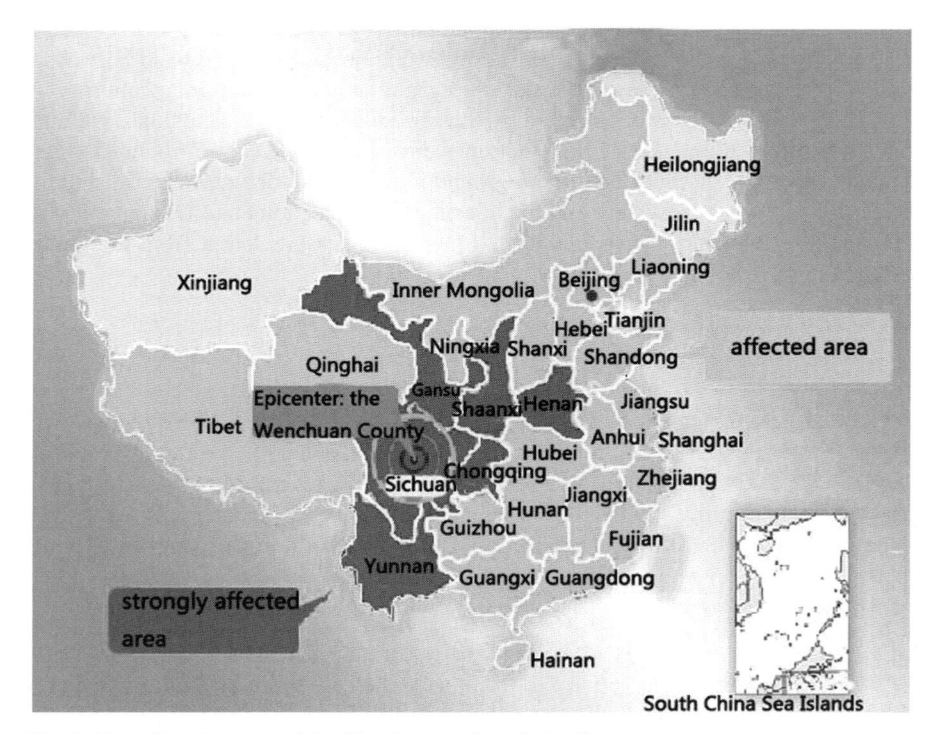

Fig. 1 The affected regions of the Wenchuan earthquake in China

Fig. 2 The Beichuan county, destroyed by the Wenchuan earthquake

within one week in 1976. The fault where the Great Wenchuan Earthquake occurred is located at the southern part of the famous south-north seismic belt in China, an no precursor in formation was observed by the newly completed China Geophysical and Geochemical Network. (Wang, 2008).

The earthquake affected large areas, which includes 30 provinces and municipals in China. The affected regions are shown in Fig. 1. The seriously damaged areas include Sichuan, Gansu and Shaanxi Provinces. The earthquake had an extremely high intensity over a large affected area with sustained impact, and the epicentral intensity is XI on Chinese Intensity Scale at the Beichuan County and Yingxiu Town of Wenchuan County. The Beichuan County and Yingxiu Town destroyed by the great event are shown in Figs. 2 and 3. The earthquake intensity zone is shown in Fig. 4, provided by the Chinese Earthquake Administration (CEA). As of October 21, 2008, a total of 34,417 aftershocks had been observed (CENC, 2008), which include 8 earthquakes measuring Ms6.0 to Ms6.9, 32 events between Ms5.0 and Ms5.9, and 231 events between Ms4.0 and Ms4.9. Figure 5 shows the distribution of aftershocks with magnitude above 4, including the main shock of the earthquake.

2 The Characteristics and Distribution of Ground Motion Acceleration of the Main Shock

The China Strong Motion Networks Center (CSMNC) was begun to constuction in 2003, and fortunately opened for operation in March, 2008, 2 months before the Wenchuan Earthquake occured. The distribution of CSMNC is shown in Fig. 6.

Fig. 3 The Yingxiu town, Wenchuan county, destroyed by the Wenchuan earthquake. (From news.sohu.com)

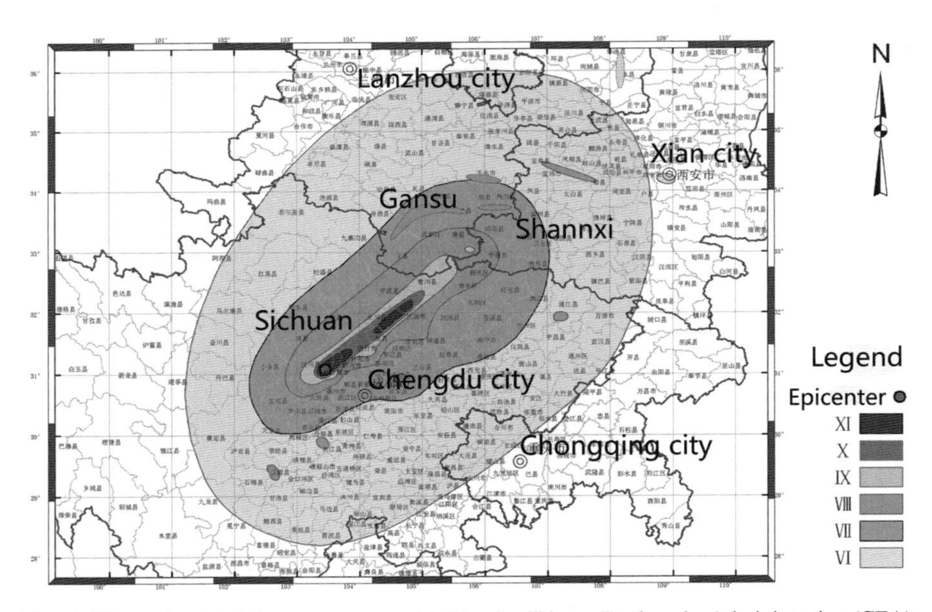

Fig. 4 The earthquake intensity zone, provided by the Chinese Earthquake Administration (CEA) (See also Color Plate 18 on page 216)

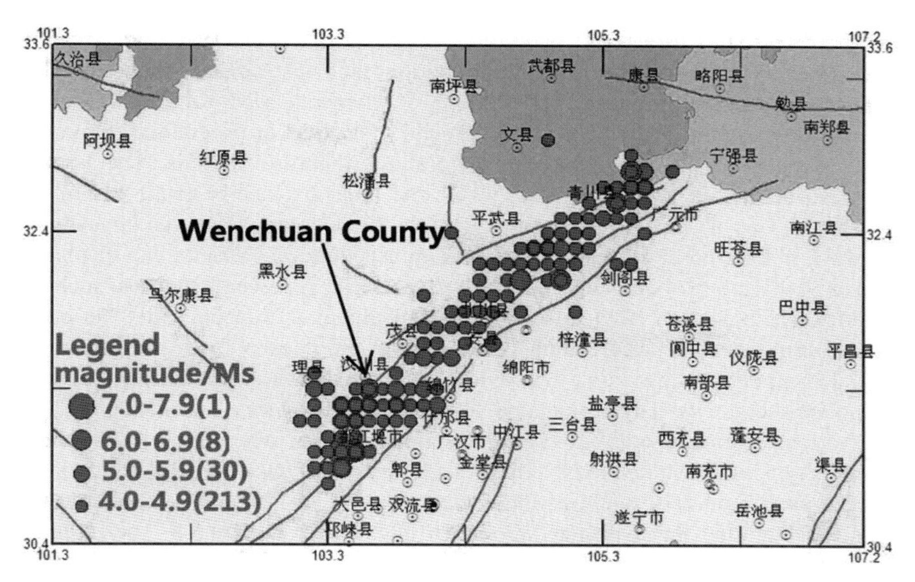

Fig. 5 The distribution of aftershocks with magnitude above 4 (From the China Earthquake Networks Center (CENC))

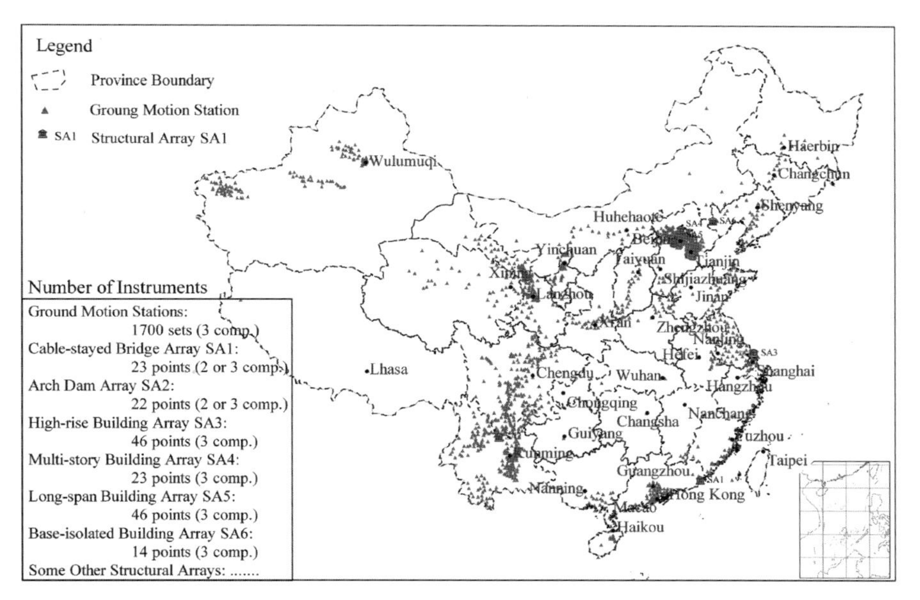

Fig. 6 The distribution of the China Strong Motion Networks Center (CSMNC)

There are hundreds of stations of strong ground motion located in the quake-hit areas. However, tens of the observation stations and instruments in the heavily quake-hit area were damaged, 460 stations of them recorded valuable information and there were about 1,400 strong ground motion records acquired. Figure 7 shows the stations that observed the main shock records in China. The largest one was obtained at Wolong station, where the fault distance is about 23 Km, in Dujiangyan City as shown in Fig. 8. Among the three components, the largest one was the EW horizontal one, and the maximum acceleration, velocity and displacement is 959.1 gal, 50.1 cm/s and 12.7 cm respectively. At several stations, such as Bajiao town in Shifang City and Jiuzhaizhuangzhatai station etc., that the peak vertical component of acceleration is larger than the peak of both the two horizonatal components, providing new evidence in the near-sourse area that vertical components could be larger than the horizontal components(Wang, 2008). The observation contradicts current seismic design guidelines, where the vertical component is typically assumed to be a third of the horizontal ones. The distributions of strong ground motion of the horizontal components (EW, NS) and vertical components (UD) around the epicenter are shown in Figs. 9, 10, and 11. The number of obtained strong motions records above 10 gal for the main shock is listed in Table 2.

Through preliminary analysis on these records, some conclusions are presented as following. (1) The distribution of PGA accords with distribution of seismic intensity well, see Fig. 12. (2) That the PGA which recorded at the stations where locates at the direction of the fault rupture are relatively greater than the others, and

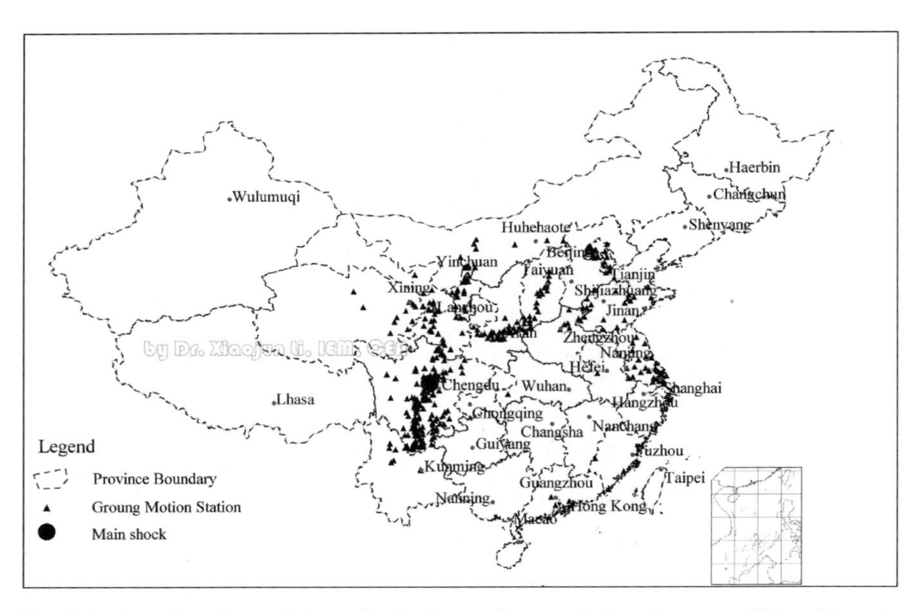

Fig. 7 Stations that observed the main shock records around the epicenter in China (From the China Strong Motion Networks Center (CSMNC))

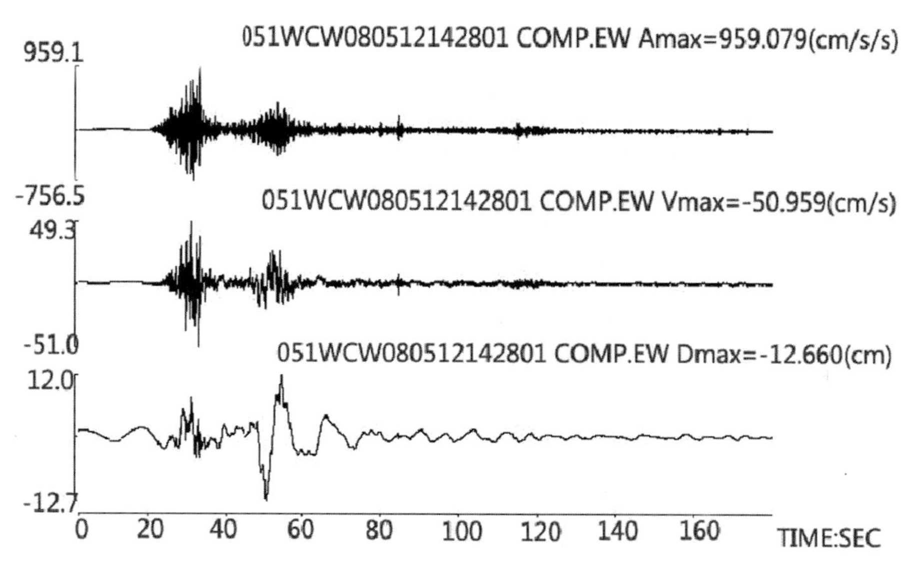

Fig. 8 The ground strong motion obtained at Wolong station in Dujiangyan city (From the top to bottom are acceleration, velocity and displacement respectively.) (From the China Strong Motion Networks Center (CSMNC))

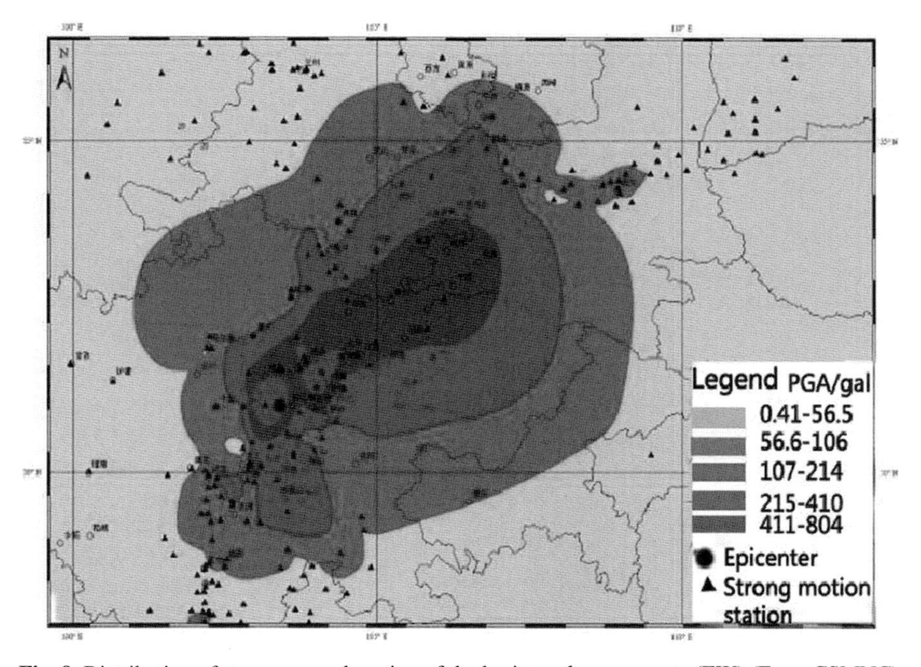

Fig. 9 Distribution of strong ground motion of the horizontal components (EW) (From CSMNC)

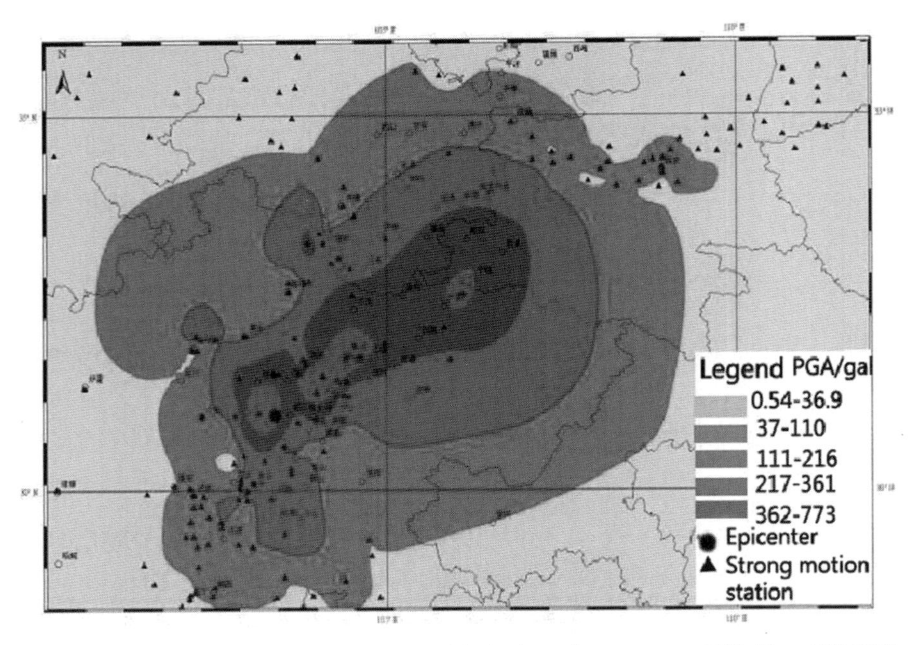

Fig. 10 Distribution of strong ground motion of the horizontal components (NS) (From CSMNC)

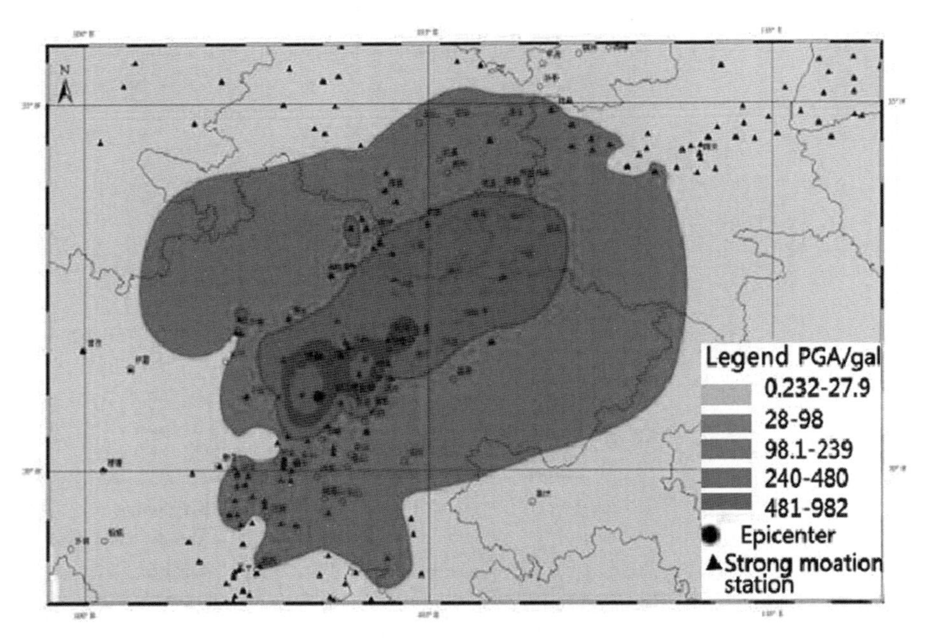

Fig. 11 Distribution of strong ground motion of the vertical components (UD) (From CSMNC)

Table 2 The number of obtained strong motion records above 10 gal

Intensity	< VI	VI	VII	VIII	IX	≥ X
Component (gal)	10.0–44.9	50.0–89.9	90.0–177.9	178.0–353.9	354.0–707.9	≥ 708.0
EW	124	31	35	14	5	2
NS	133	28	36	15	4	1
UD	86	30	12	7	5	1
Total number	343	89	83	36	14	4

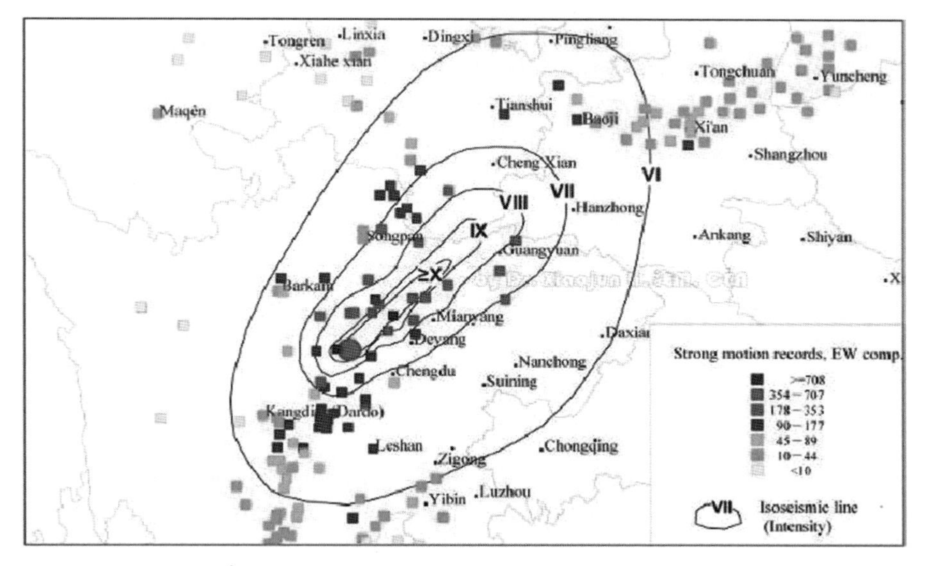

Fig. 12 The distribution of EW component records (From prof. Li Xiaojun)

(3) the vertical components could be larger than the horizontal components in the near-source area.

3 Characteristics of Damage of Buildings and Houses

There are five kinds of buildings and houses in the quake-hit area, the Sichuan, Gansu and Shaanxi Province, which includes reinforced concrete buildings, brick-concrete buildings, the bottom reinforced concrete buildings, brick-wood houses and adobe-wood houses. According to field investigation on buildings and houses damage in the Dujiangyan City, Sichuan Province, The reinforced concrete buildings were damaged slightly, and then the bottom reinforced concrete buildings took the second place. The buildings and houses built after 2000 were damaged much more slightly than the old ones, because many of them were constructed according the National Building Seismic Design Code (2001). The damage proportion for

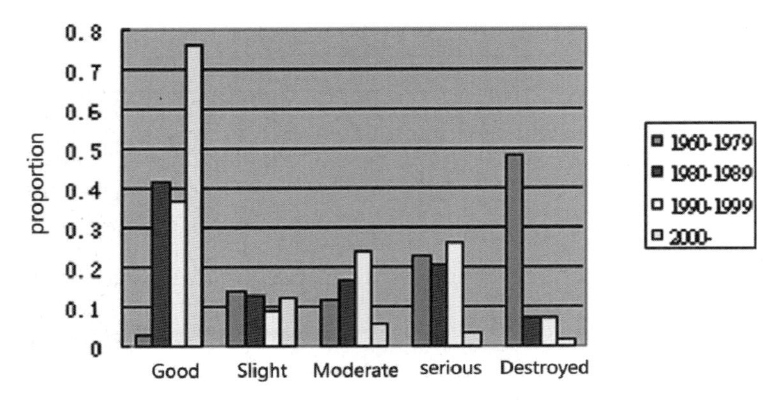

Fig. 13 Damage proportion for buildings built in different years (From Zhang Minzheng)

buildings built in different years is shown in Fig. 13. The characteristics of damage of these buildings and houses are described as following.

3.1 Reinforced Concrete Buildings

The reinforced concrete buildings mainly locate in the cities and towns of the heavily affected area, and which were proper seismic design and high-quality construction. Even in towns with almost complete destructions, there were always a few RC buildings that did not suffer heavy damage, except the buildings located at a place where a fault rupture crossing. Figure 14 shows a standing building in the Dongfang Steam Turbine Works located in Hanwang town, which was damaged during the

Fig. 14 A standing building in the Dongfang Steam Turbine works located in Hanwang town

Fig. 15 A business building in Hangwang town

earthquake. Field investigation proved that the first and second floors were sheared damage to the walls, however the beams and columns did not damaged and the structure could be used after repairs. Figure 15 shows a business building in Hangwang town. Note the walls of this building were collapsed but the frame kept in good situation. Most of the reinforced concrete buildings have a good performance during the great earthquake shocking, in which the main frame did not collapse.

3.2 Bottom Reinforced Concrete Buildings

The bottom reinforced concrete buildings are also locate in cities and towns of the quake-hit region. In general, the first floors were constructed by reinforced concrete and the upper floors were brick-concrete contracture. Figure 16 shows a five floors building in Dujiangyan City. The second floor was sheared to damage and the upper

Fig. 16 A five floors building in Dujiangyan city

Fig. 17 The building with the top floor destroyed in Guangji town, Mianyang city

floors were minor damage to the walls. Figure 17 shows a building which the top floor was destroyed in Guangji Town of Mianyang City.

3.3 Brick-Concrete Buildings

The brick-concrete buildings have low seismic resistance, especially the buildings with precast slabs. The intensities were 10 and 11 at the heavily affected area, which was much higher than specified by the building design codes for Wenchuan and Beichuan County. Therefore, 90% of the brick-concrete buildings collapsed or were heavily damaged in Sichuan Province. A great number of people was died due to the collapses of brick-concrete buildings with precast slabs. Figure 18 shows the complete collapse of a school building in Yingxiu Town of Wenchuan County. Figure 19 shows a building which the first floor was destroyed and the upper floors heavily damaged in Beichuan County.

Fig. 18 Completely collapse of a school building in Yingxiu town of Wenchuan county. (From Gao Mengtan)

Fig. 19 A building in Beichuan county, which the first floor was destroyed and the upper floors were heavily damaged

Fig. 20 The weather bureau of Longnan City in Wudu district

In the heavily quake-hit area in Gansu Province, which the intensities were 8 and 9, most of the brick-concrete buildings kept standing but were damaged. The upper floors were damaged worse than the floors below. Figure 20 shows a building of the Weather Bureau of Longnan City in Wudu District, where the intensity was 8. The top floor of the building was sheared to damage badly, and this building was removed later.

3.4 Brick-Wood Houses and Adobe-Wood Houses

The brick-wood houses and adobe-wood houses locate in the rural area in Sichuan, Gansu and Shaanxi Province. The current design code does not cover buildings in rural areas in China, making these buildings vulnerable when an earthquake occurs. These kinds of houses have a very low level of earthquake resistance and most of them were built before 2000. Almost all of these houses collapsed in the region with

Fig. 21 Collapsed houses in a village located in Jiulong town, Mianzhu city

Fig. 22 The completely destroyed village of Haoping village in Wudu district, Longnan city

intensity of 10 and 11, causing a huge number of casualties and loss as shown in Fig. 21. In the area with intensity of 8 and 9 of Longnan City in Gansu Province, The brick-wood houses and adobe-wood houses which located at the top mountain or mountainside were collapsed. Figure 22 shows the completely destroyed village of Haoping village in Wudu District of Longnan City.

4 The Characteristics of Damage To Infrastructures and Lifeline

In the heavily quake-hit area, many types of infrastructures and lifeline engineering structures suffered damages to different extents, such as highway and railway, bridge, dam, electricity supply, water and gas supply, and communication system.

4.1 Highway and Railway

Many highways were destroyed by the main shock and the secondary disasters induced by the great earthquake, such as landslides and rolling rocks. Figure 23 shows the Duwen highway connecting the Dujiangyan City and the Wenchuan County, which was opened for traffic on May 11, 2008 and destroyed by the main shock completely on May 12, 2008. It had only operated for 1 day. Figure 24 shows a subside with 10 m long and 3 m wide in the G212 highway, which connects Gansu Province and Sichuan Province in Wenxian County, Gansu Province. Moreover, the earthquake caused damage to railways. The railway track deformed and a freight train overturned at the Yinghua Town, Shifang City. No.109 tunnel of Baocheng railway, which connects the Baoji City and the Chengdu City, was damaged by a landslides caused by the earthquake as shown in Fig. 25. A train contained airplane gasoline was getting across the tunnel when the main shock occurring. The train had fired and stopped the traffic of the railway a few days.

Fig. 23 The Duwen highway, which connects the Dujiangyan city and the Wenchuan county, was completely destroyed

Fig. 24 A subside with 10 m long and 3 m wide in the G212 highway

Fig. 25 No.109 tunnel of Baocheng railway was damaged by a landslides caused by the earthquake (From Du Zezhong)

4.2 Bridge

There were more than 400 bridges damaged by the huge earthquake, which included simply supported beam, steel truss, suspension and arch bridges. Some of them were collapsed or seriously damaged in the turn section and the rupture zones due to failure or dislocation of the piles and foundation. The combination effect of ground surface rupture and main strong shaking caused the damage to bridges. Figure 26 shows the damage to Xiaoyudong Bridge, which was 187 m long and operated for traffic in 1999. A few bridges were damaged by after-shocks. Figure 27 shows a bridge in Wenxian County, which was completely destroyed by the after-shock with magnitude of 6.1, occurred on August 5.

Fig. 26 Damage to the Xiaoyudong bridge

Fig. 27 The Luoxuangou bridge, which connects the Wenxian county of Gansu province and the Guanyuan city of Sichuan province, in Wenxian county

4.3 Dam

The damage to dams was not serious, that could seem to be due to the big dams were designed for a seismic intensity of 8 in the quake-hit area. Zipingpu reservoir is one of the largest reservoirs along the Mingjiang River, which has a rock-fill dam with a concrete cover plate, and the height of it is 156 m and length 663 m. The hydropower plant has been working well since the main shock occuring. Figure 28 shows a settlement of 30 cm at the top the dam. Figure 29 shows the Bikou reservoir in Wenxian County, which was damaged slightly by the earthquake.

Fig. 28 The settlement of 30 cm at the top the dam of the Zipingpu reservoir

4.4 Power, Water and Gas Supply

The power supply was stopped in the seriously quake-hit area at the several beginning days of the main shock. More than 100 transmission towers were damaged. Some transmission towers and power substations were found to be damaged as shown in Figs. 30 and 31. Water and gas supply was not seriously damaged during our survey. Figure 32 shows a water supply pipe was broken after the earthquake.

Fig. 29 Damage to the Bikou reservoir in Wenxian county

Fig. 30 A power transmission tower was damaged. (From Xinhuanet, 2008)

Fig. 31 A power substation was damaged in Yingxiu town, Wenchuan county

Fig. 32 A water supply pipe was broken after the earthquake (From Yuan Yifan, 2008)

4.5 Communication System

All communication with the outer world of the seriously quake-hit area had been stopped a few days by the main shock. Even in Lanzhou City, 505 km away from the epicenter, the communication was congested in the first time and stopped about 1 hour. Figure 33 shows a set of communication cables was pulled down.

Fig. 33 A set of communication cables was pulled down

5 The Geotechnical Disasters Induced by the Earthquake

5.1 Faults Rupture

The great earthquake occurred along the Longmen fault locating at the southern part of the famous south-north seismic belt, which has an up bound magnitude of 7.3 for potential seismic sources on the zonation map in China (Wang, 2008). The main ground rupture is more than 200 km long through the science survey

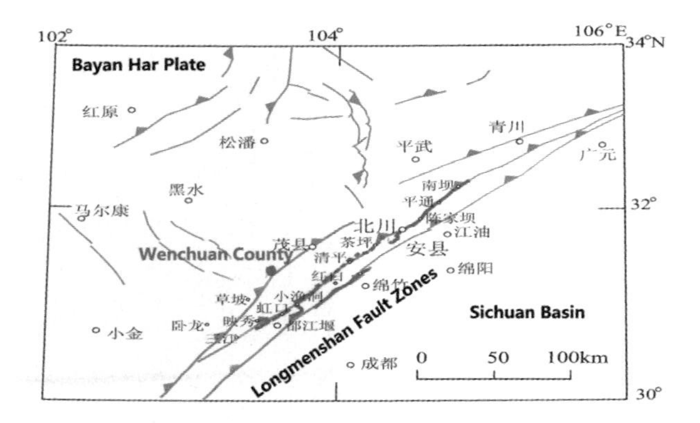

Fig. 34 Spread of the ground rupture (*red line*) caused by the earthquake

for the earthquake organized by CEA, and the secondary rupture is about 50 km in length. The maximum vertical dislocation is 6.4 m and the horizontal dislocation 5.5 m respectively. The fault is a left lateral thrust one, and the north-west plate uplifted during the main shock. Figure 34 shows the spread of the ground rupture of the earthquake. Figure 35 shows the uplifted of a county road in Hongkou, Dujiangyan City. Figure 36 shows a rupture crossed between two buildings of the Bailu high school. During the survey in the quake-hit areas (see Fig. 37), it was found that the fault rupture completely destroyed buildings and houses within 50 m, seriously damaged within 100 m, medium damage within 150 m and kept in good condition beyond 200 m. The fault rupture crossed the Xiaoyudong Town, the relationship between destructive degree of buildings and fault distance is shown in Fig. 38.

5.2 *Quake Lake*

There were 34 Quake lakes found in the affected area, because of rocks plugged the Qingshui River and its branch, the Hongshi River. The largest one is the Tangjia

Fig. 35 Uplifted of a county road in Hongkou, Dujiangyan city

Fig. 36 A rupture crossed between two buildings of the Bailu high school

Fig. 37 The fault rupture crossed the Beichuan county. (Towards to SW)

Fig. 38 The fault rupture crossed the Xiaoyudong town

Mountain quake lake, which had a dam with 82.8 m in height and 220 m in length. The quake lake could flood more than 1.3 million people lower downstream, including people in the Bechuan County, the Shifang and Deyang city etc. The photos before and after the quake lake formed are as shown in Fig. 39.

Fig. 39 The photos before and after the quake lake formed (Taking time: the *left*, 2006-05-14 (the *right*, 2008-05-22)) (From Xie Lili, 2008)

5.3 Landslides, Rolling Rocks and Debris Flow

Landslides, rolling rocks and debris flow were the three main types of secondary geotechnical damage. Many places were seriously affected by huge landslides, slope collapses, rolling rocks and debris flows after the main shock. It caused the destruction of constructions, infrastructures, buildings and houses, which brought a great number of deaths and injuries. More than 7,000 people were buried by huge landslides in Beichuan County (see Fig. 40) induced by the main shock. Figure 41 shows rolling rocks, which crashed a bus and caused 10 people died in Wenxian County,

Fig. 40 More than 7,000 people were buried by the huge landslides induced by the main shock in Beichuan county

Fig. 41 Rolling rocks crashed a bus and caused 10 people died in Wenxian county, Gansu province

Fig. 42 A debris flow buried a village in the Jiufeng mountain, Pengzhou city

County, Gansu Province. Figure 42 shows a debris flow buried a village in Jiufeng Mountain, Pengzhou City.

5.4 Liquefaction and Settlements

There were total 38 liquefied sites found in the quake-hit area of Sichuan and Gansu Province. Liquefaction caused damage widely to houses, buildings, farmlands, fish pounds, irrigation channels, underground wells, bridges, and roads etc. It was observed that liquefaction developed in sand and gravel, and some of them were sand with medium size. Some buildings were damaged due to liquefaction. It is unexpected that liquefaction was developed in silt and gravel deposits in the sites with low earthquake intensities. Figure 43 shows a liquefied site in a farmland in Wudu District, Longnan City, Gansu Province. Some seismic settlements were found in losses regions of Gansu Province. Figure 44 shows the seismic settlements occurred at a farmland in Qingshui County, where the intensity was 6.

Fig. 43 A liquefied site in a farmland in Wudu district, Longnan city, Gansu province

Fig. 44 Earthquake settlements occurred at a farmland in Qingshui county, where the intensity was 6

6 Lessons Learnt from the Great Wenchuan Earthquake

Some lessons learnt from the earthquake were summarized as follows: (1) The collapses of brick-concrete buildings with precast slabs was the first reason that caused the biggest number of deaths and injuries by the huge earthquake. This kind of buildings was not built according to current seismic design. However, most of buildings with well seismic design and good construction quality indeed did not damage even though encountering much stronger ground motion as expected (see Figs. 45 and 46). (2) Most of the reinforced concrete buildings, in the seriously affected area, had a good performance during the Earthquake, in which the main frame did not collapse. (3) Large-span rooms and staircases were the weak parts of buildings, many of them collapsed. The construction idea of "Strong columns and weak beams" for houses and buildings didn't been come true in the affected areas. Actually, most of columns of the buildings in the quake-hit area damaged first, instead of beams. (4) Many bridges were collapsed or seriously damaged in the turn section and the rupture zones due to failure or dislocation of the piers and

Fig. 45 The new village in Wudu district, Longnan city, Gansu province

foundation. (5) Landslides were the typical secondary disasters caused more than 8,000 people died. The new sites for reconstruction in villages and towns should be evaluated in terms of seismic safety. (6) The effect of site amplification on damage was remarkable at the sites at deep loess and the top of mountains, which was observed in Gansu Province. The higher the landform, the worse the quake disaster. The earthquake intensity at the top of mountains is about 1 degree higher than that of valleys. Figure 47 shows the different displacements of wood columns of two adobe-wood houses, which locate at the same mountain and very close. The left house locates at the top of mountain, and the right one at the valley, which is about 150 meters lower than the left one. The displacement of the left one was 5 cm, the right 1 cm. The deeper layer of loess, the more serious the damage. Figure 48 shows a site went against anti-seismic at the Wudu District. The buildings which locate at the top of the hill were damaged seriously than which at the bottom. The Qingyang City of Gansu Province, where the losses layer is very deep, 669 km away from the epicenter of the huge earthquake. However, the damage to the city was serious. Figure 49 shows a loess cave-house, which locates at the Xifeng District, Qingyang

Fig. 46 A new village in Huating county, Gansu province

Fig. 47 Different displacements of wood columns of two adobe-wood houses in Haoping village, Wudu district, Longnan city

Fig. 48 A site at the Wudu district, Longnan city

City, was collapsed by the earthquake. This is one of the farthest damaged buildings observed from the epicenter. (7) It is unexpected that liquefaction was developed in silt and gravel deposits in the sites with low earthquake intensities.

Fig. 49 A loess cave-house, which locates at the Xifeng district, Qingyang city

7 Post-Disaster Reconstruction

By December, 2008, a fund of 70 billion Chinese Yuan, invested by the central government of China, has been used for reconstruction in the seriously disaster areas, such as the Sichuan, Gansu and Shaanxi Province. The local governments have invested lots of money and material for reconstruction as well. Moreover, several policies and office procedures were made for instructing and supervising construction.

In order to improve the seismic design of buildings and structures in the seriously quake-hit areas, the zonation map of ground motion parameter of the Wenchuan Earthquake was presented, on the base of field investigation, revising the national zonation map of ground motion parameter, and researching on seismogeology and circumstance of warning signs of earthquake, activity of the Longmen Mountain belt, potential focal region and recurrence period of great earthquake. Figures 50 and 51 show the zonation map of PGA and characteristic period of ground motion response spectrum of the seriously damaged area respectively. Also, the distribution map of active faults in the seriously affected area by the great earthquake was made for post-disaster reconstruction. The newly built buildings and structrues must be kept away from the active faults. Figure 52 shows the active faults distribution in the south-east area of Gansu Province, where is the seriously damaged area in this province.

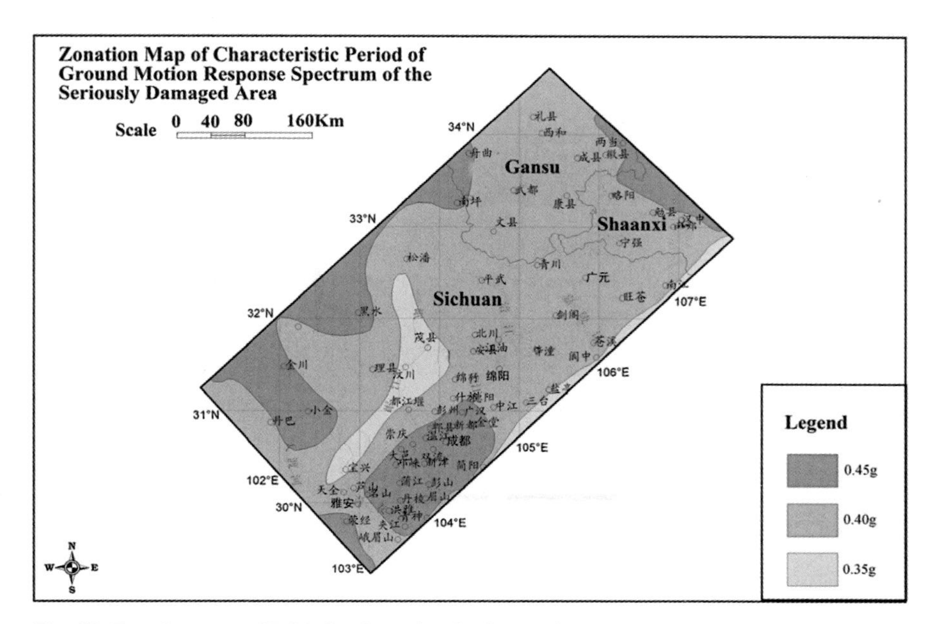

Fig. 50 Zonation map of PGA for the seriously damaged area (See also Color Plate 19 on page 216)

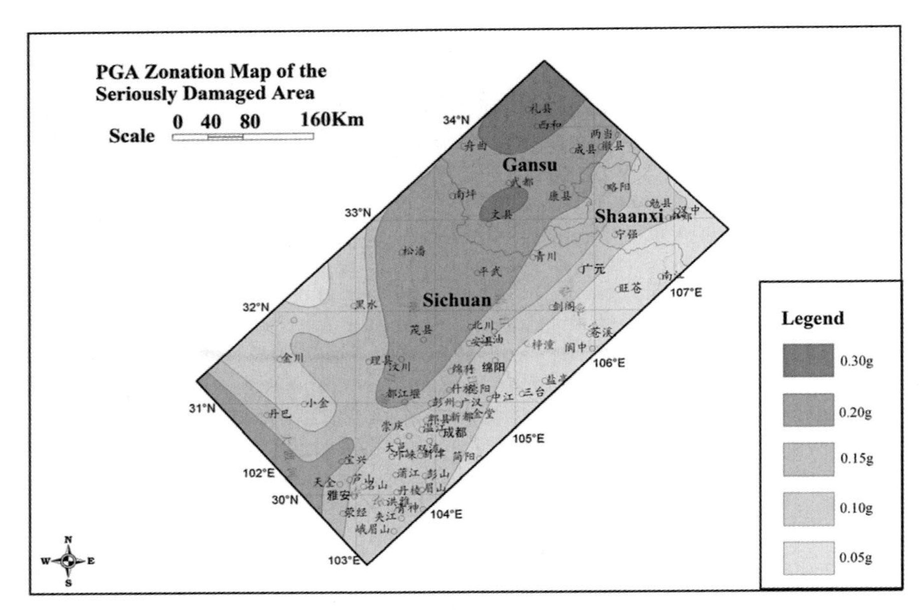

Fig. 51 Zonation map of characteristic period of ground motion response spectrum for the seriously damaged area (See also Color Plate 20 on page 216)

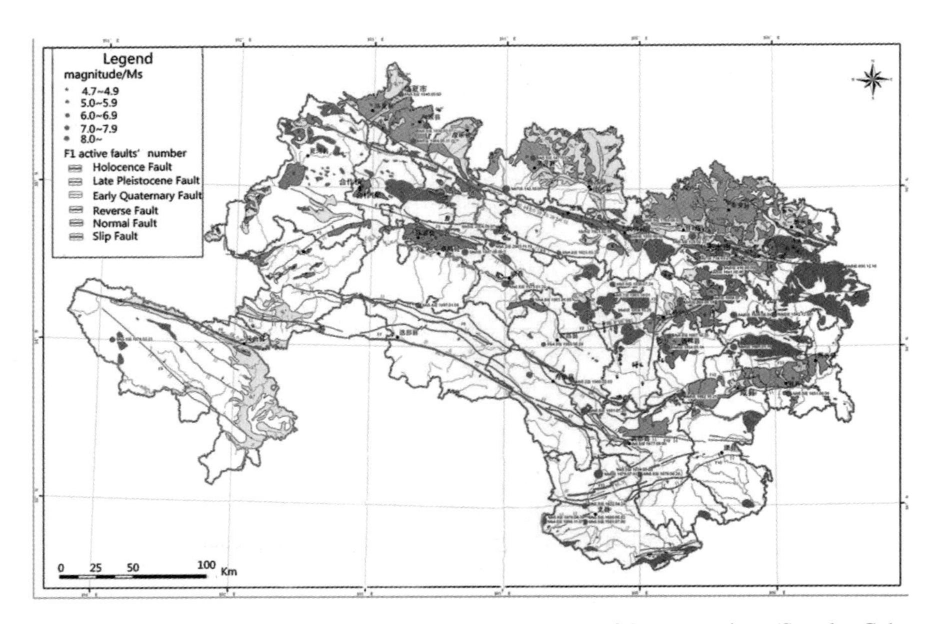

Fig. 52 Map of active faults distribution in the south-east area of Gansu province (See also Color Plate 21 on page 216)

Fig. 53 A newly built house for post-disaster construction in the Kangxian county of Gansu province

Fig. 54 A newly built village in the Wudu district of Longnan city, Gansu province

According to the guideline for seismic design of buildings, many measures should be taken for improving the ability of anti-seismic in quake-hit regions, such as ring beam, structure column, and etc. Figure 53 shows newly built house for post-disaster construction in the Kangxian County of Gansu Province, and Fig. 54 shows a newly built village in the Wudu District of Longnan City, Gansu Province.

8 Conclusions

The most efficient and reliable methods, which are currently available for relieving the casualties and loss caused by earthquake, are reasonable seismic prevention, formal seismic design and good construction quality for buildings and infrastructures. The buildings and infrastructures designed according to the present Chinese code for seismic design of buildings had obviously better performance under the effects of the Wenchuan great earthquake than those without seismic design.

Among buildings with various kinds of structures, reinforced concrete buildings had the best seismic performance, and then the bottom reinforced concrete buildings, brick-concrete buildings, the brick-wood houses respectively and adobe-wood houses had the worst performance.

Most of death toll was caused by both Collapses of brick-concrete buildings and geotechnical disasters, especially the landslides and rolling rocks induced by the earthquake.

Seismic capability of houses, public infrastructures for both farmers and town citizens should be improved in China. Sites for construction in buildings and houses should be evaluated in terms of seismic safety.

The effect of site amplification on ground motion was remarkable in terms of the field damage investigation.

The fault rupture completely destroyed most of the buildings and houses within 50 m, seriously damaged within 100 m, medium damage within 150 m and in good shape beyond 200 m.

Acknowledgments The authors thank to Prof. Xie Lili, Prof. Yuan Yifan and Prof. Wu Yaoqiang for their providing some related data and help in the field investigation. The paper is supported by The Special Fund of The State Social Commonweal (2004DIB3J130) and The Basic Research Fund of The Earthquake Science Institute, CEA.

References

China Earthquake Networks Center (CENC), 2008, China Earthquake Networks Center Data.
Wang, Z.F. 2008. A preliminary report on the great Wenchuan earthquake. Earthquake Engineering and Engineering Vibration, Vol.7, No.2.

Color Plates

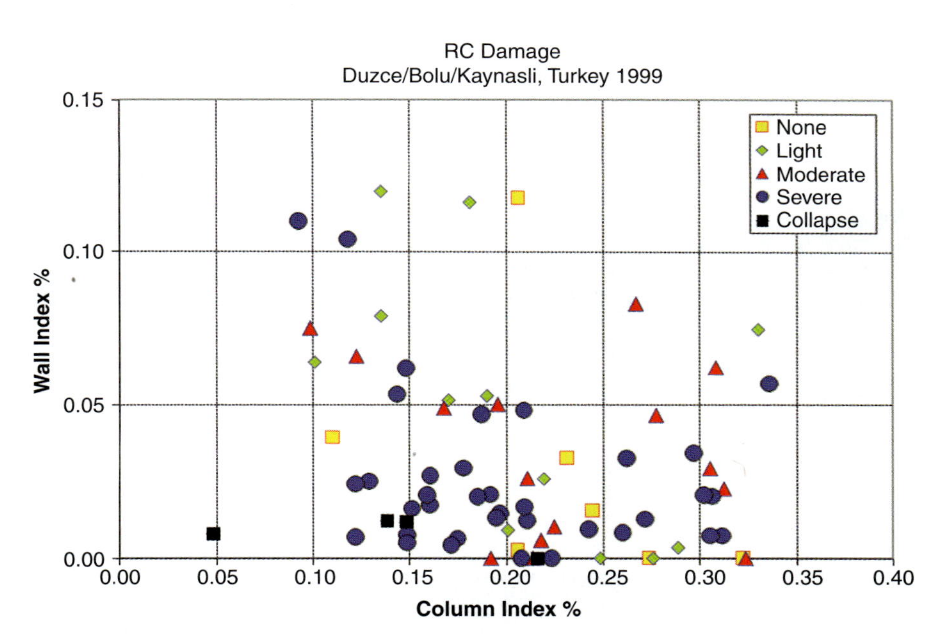

Plate 1 (See also Fig. 2 on page 3)

Plate 2 Fracture of CFRP sheet in Beam 2 (See also Fig. 8 on page 27)

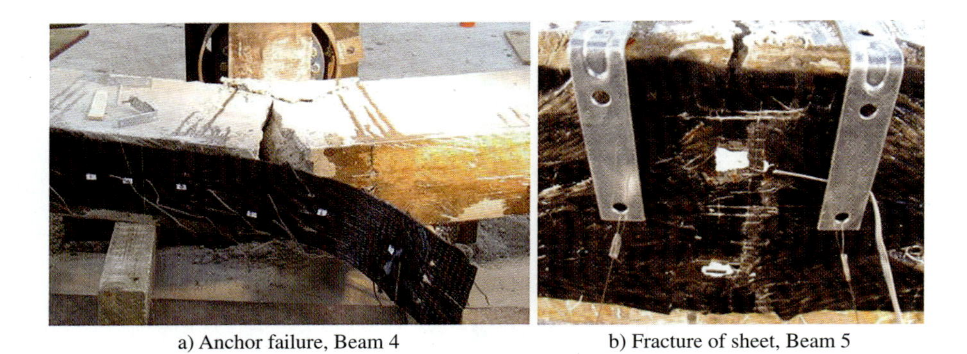

a) Anchor failure, Beam 4 b) Fracture of sheet, Beam 5

Plate 3 Failure with no adhesion between CFRP sheet and concrete (See also Fig. 9 on page 27)

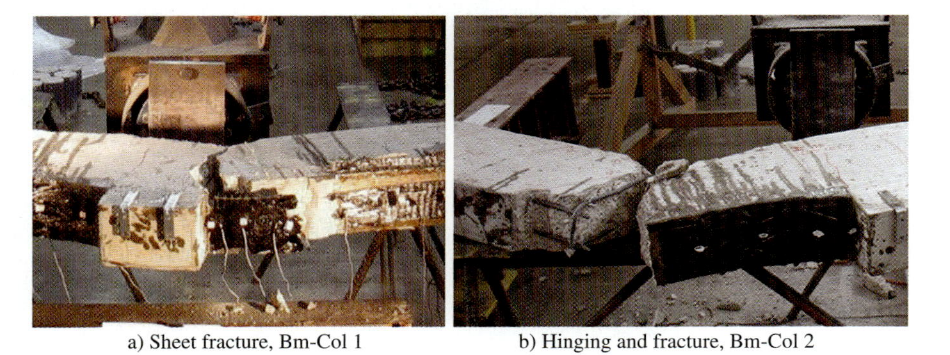

a) Sheet fracture, Bm-Col 1 b) Hinging and fracture, Bm-Col 2

Plate 4 Failures of Beam-Column specimens (See also Fig. 10 on page 28)

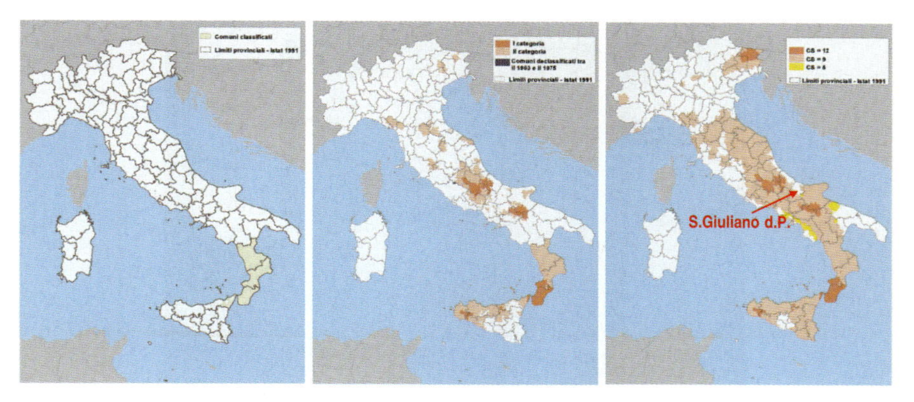

Plate 5 Historical evolution of seismic classification in Italy (De Marco and Martini 2001) (See also Fig. 4 on page 72)

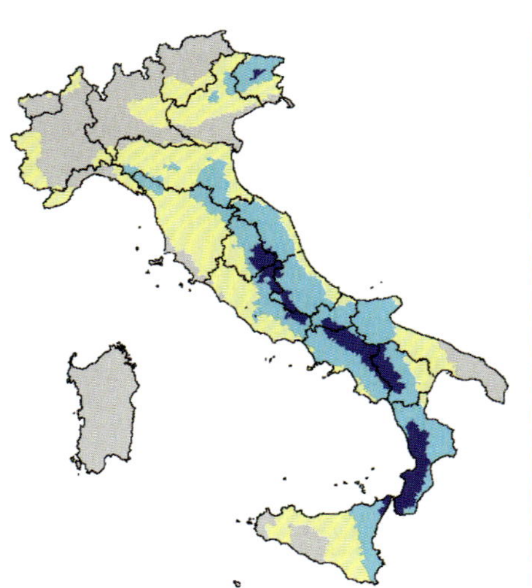

Plate 6 Seismic classification proposal of 1998 (Gavarini et al. 1999) (See also Fig. 5 on page 72)

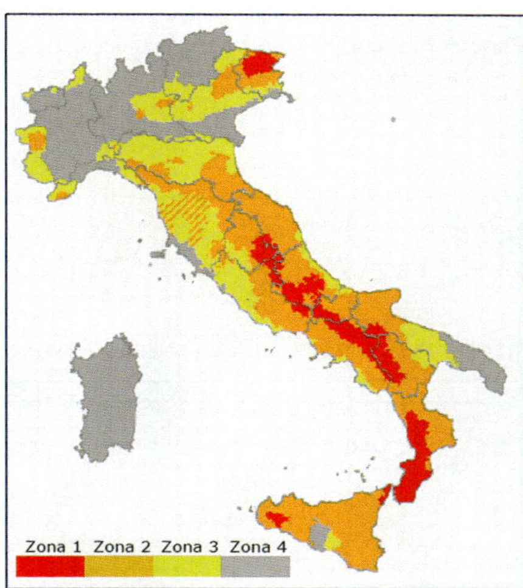

Plate 7 Current Italian seismic zonation as determined by the by-law 3274/2003 (See also Fig. 7 on page 76)

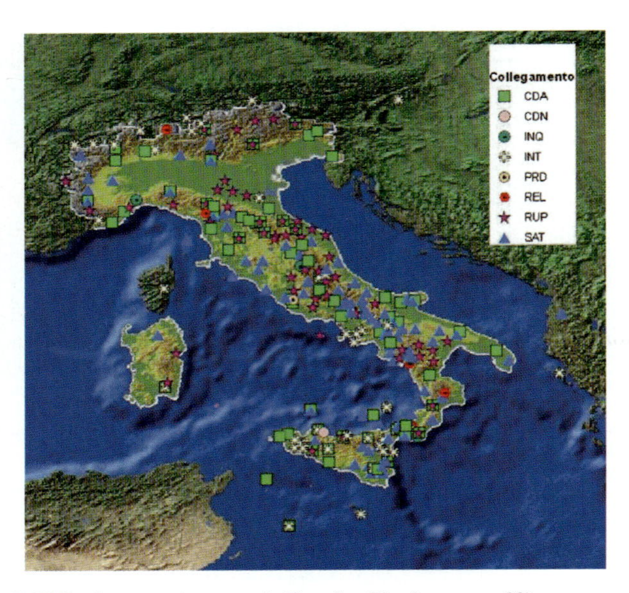

Plate 8 Italian INGV seismometric network (See also Fig. 9 on page 83)

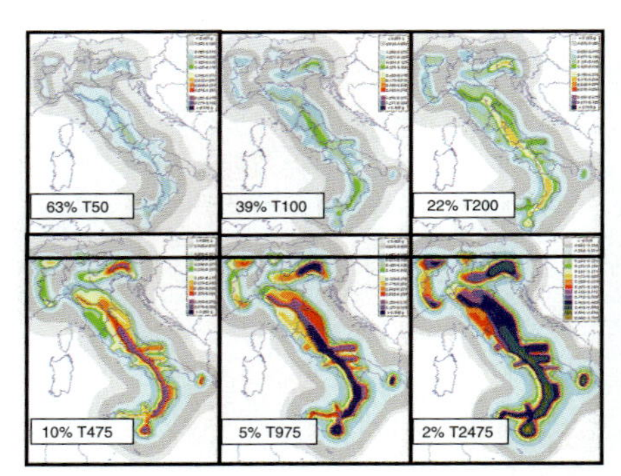

Plate 9 Hazard maps in terms of peak ground acceleration on stiff soil (ag) for different return periods (project DPC-INGV-S1) (See also Fig. 10 on page 84)

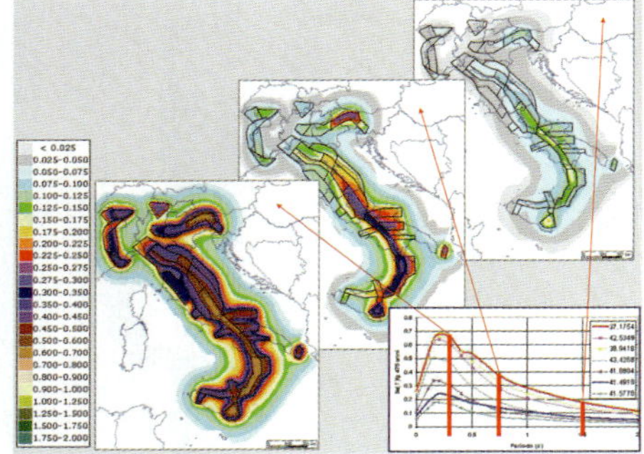

Plate 10 Hazard maps in terms of elastic response spectra on stiff soil (project DPC-INGV-S1) (See also Fig. 11 on page 85)

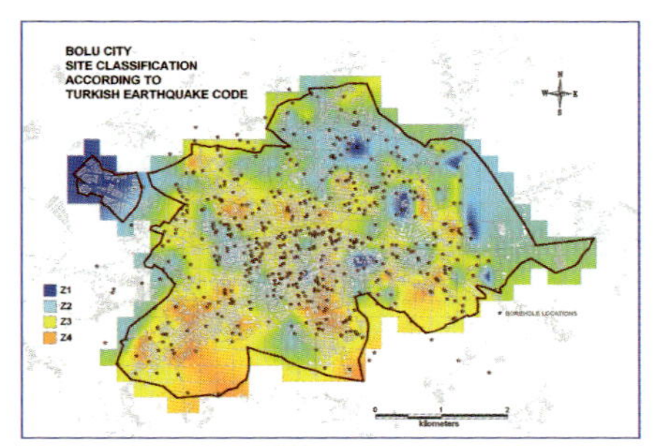

Plate 11 Microzonation with respect to Turkish earthquake code (2007) (See also Fig. 1 on page 138)

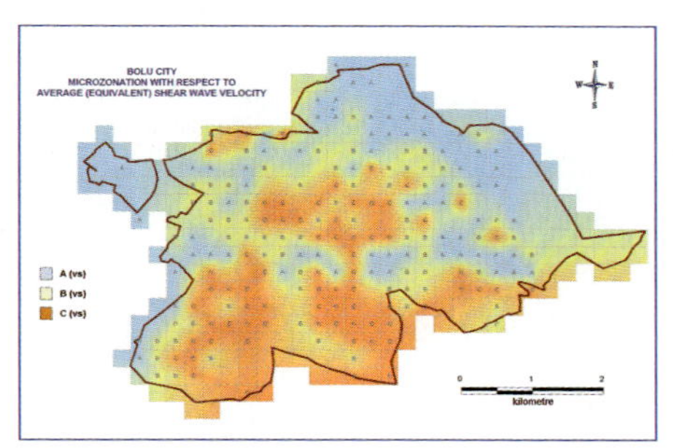

Plate 12 Microzonation with respect to average shear wave velocity (V_{S30}) (See also Fig. 4 on page 141)

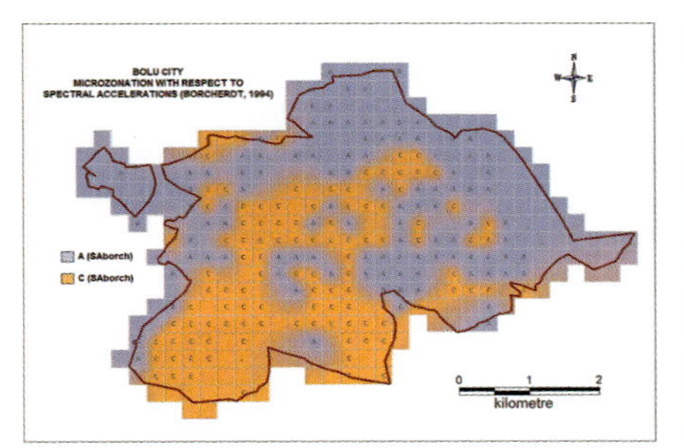

Plate 13 Zonation with respect to spectral accelerations calculated using Borcherdt (1994) procedure for Bolu city (See also Fig. 6 on page 144)

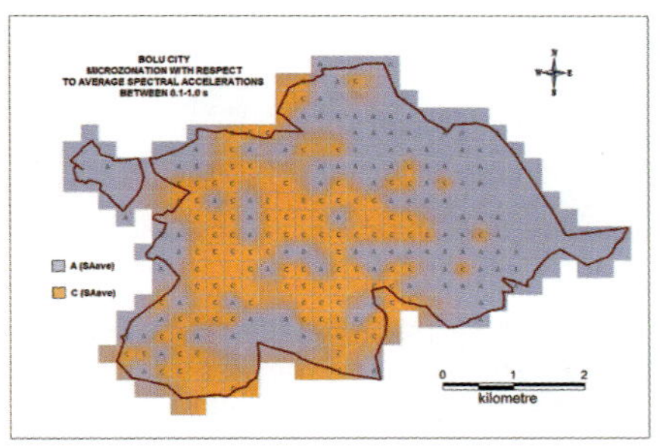

Plate 14 Zonation with respect to average spectral accelerations calculated by site response analysis for Bolu city (See also Fig. 7 on page 144)

(a) (b)

(c) (d)

(e) (f)

Plate 15 Damage to reinforced concrete frame buildings in Nai Thon Beach, Thailand (See also Fig. 2 on page 156)

Plate 16 A single-story house, displaced by water pressure due to lack of proper anchorage, Banda Aceh, Indonesia (See also Fig. 13 on page 164)

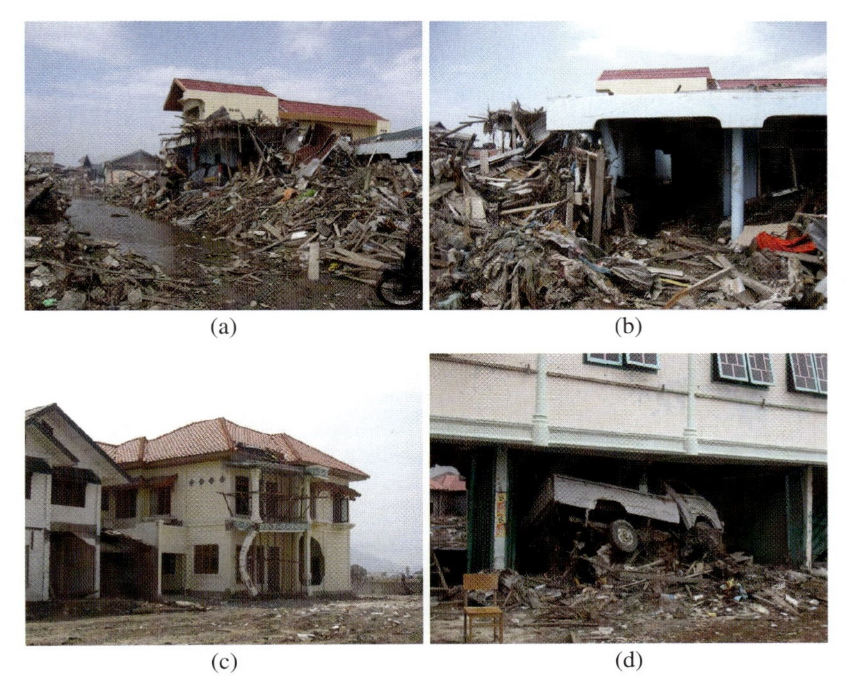

(a) (b) (c) (d)

Plate 17 Impact loading on columns due to floating debris, Banda Aceh, Indonesia (See also Fig. 15 on page 166)

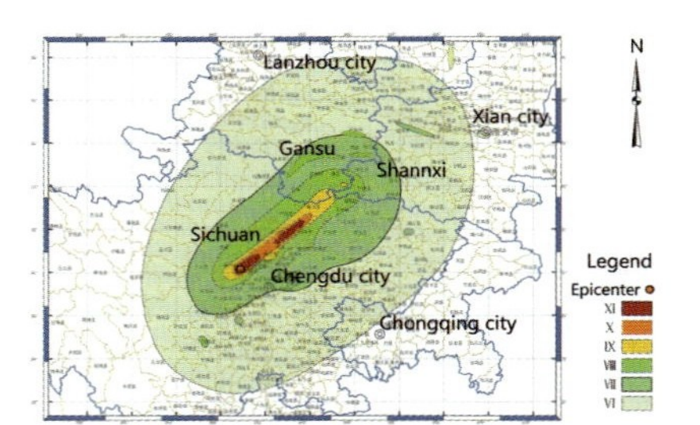

Plate 18 The earthquake intensity zone, provided by the Chinese Earthquake Administration (CEA) (See also Fig. 4 on page 182)

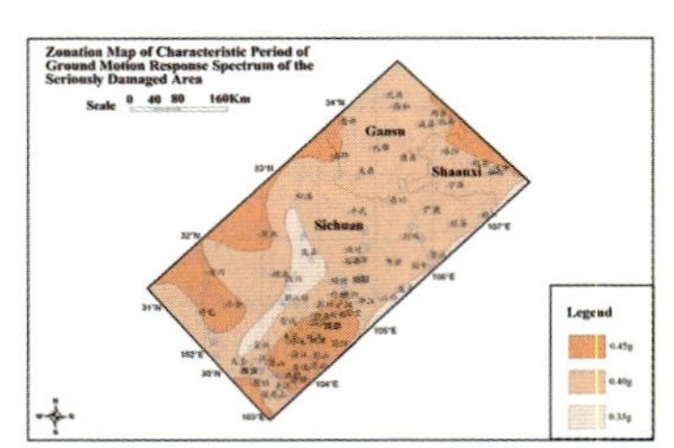

Plate 19 Zonation map of PGA the seriously damaged area (See also Fig. 50 on page 205)

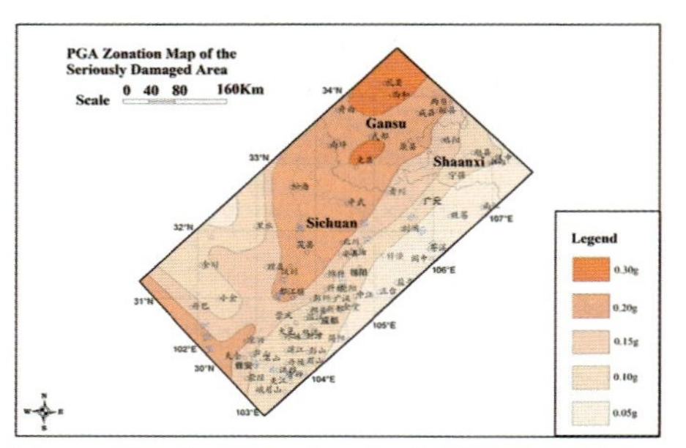

Plate 20 Zonation map of characteristic period of ground motion response spectrum for the seriously damaged area (See also Fig. 51 on page 206)

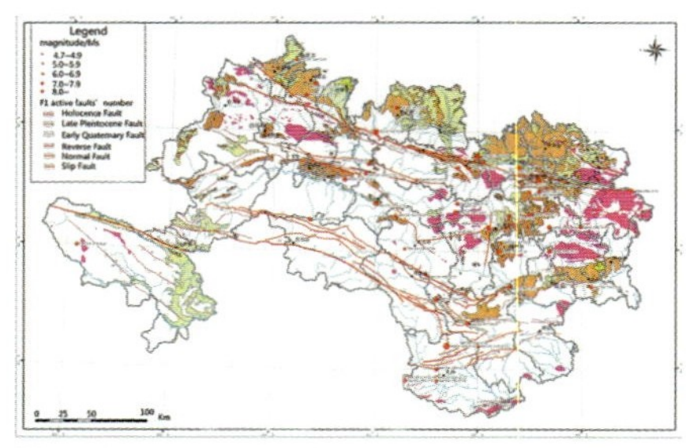

Plate 21 Map of active faults distribution in the south-east area of Gansu province (See also Fig. 52 on page 206)

Index